I0037583

From: noreply@printwhistle.com
To: You
CC: Your Boss
BCC: Your Ego
Subject: Internal Notice - Upgrade Your Career in IT

There are plenty of resources that will teach you how to pass a certification, configure some sort of technology, or give you vague and ambiguous motivational rhetoric. This book is not one of them.

This book exists to fill in the gaps. To say the quiet part out loud, point out common missteps, and shine a light on career-limiting habits that too many people in IT learn too late (or never learn at all).

It is about understanding what actually moves careers forward in this industry. If you've ever felt stuck, undervalued, or unsure of what's missing, this is your next step.

The game has always been real. You just didn't have the rulebook. Until now.

Best Regards,

Printwhistle IT Careers Division

The IT Factor: Upgrade Your Career in Technology

Copyright © 2025 by Aaron Cervasio

https://www.aaroncervasio.com

Published by Printwhistle Publishing

https://www.printwhistle.com

ISBN (Paperback): 978-1-968272-00-5

ISBN (E-Book Kindle): 978-1-968272-01-2

Published in the United States of America

Library of Congress Control Number: 2025912048

Design and Layout by Aaron Cervasio

This publication is intended to provide accurate and valuable information, based on the author's real-life experience and perspective. While every effort has been made to ensure its accuracy, the publisher and author assume no responsibility for any errors, omissions, broken links, or inconsistencies. Some names, incidents, and details have been fictionalized for clarity or confidentiality. This book is not a substitute for professional training or formal education. The views expressed are solely those of the author and do not reflect any current or past employer.

Acknowledgments

To my cousin and fellow IT professional Anthony Curtis who was hugely influential on my passion to pursue a career in technology: During our teenage years, you were a constant inspiration to me. I was always impressed—maybe even a little jealous—of your technical knowledge and drive. Thank you for being by my side in those days as a close friend and fellow aspiring technology professional.

To my mother Cheryl Cervasio, grandmother Barbara Linneman, and great aunt Dolores "Dee" Barnes, all of whom have passed on: Your love (sometimes tough) and support has helped make me the person I am today. Mom and Grandma, thank you both for helping to pay for vocational school (and many computers and computer parts), which helped me establish skills and credibility early in my career. Aunt Dee, thank you for opening your home to me and giving me a place to be me where I could use and build gaming systems. I was blessed to have you all in my corner and you are all sorely missed.

To my "Uncle" Brian Barnes: Thank you for introducing me to computers and gaming. Watching you on the computer as a young kid and joining you in playing many of those games is what kickstarted my passion for technology. I am grateful for all the late nights and summers I had sitting next to you having LAN parties.

To my dad, Brian Cervasio: You taught me what it means to be a hard-working man. You and mom instilled into me a work ethic that I carry with me to this day. Without you, I wouldn't have gotten to where I am today.

To my various colleagues over the years...

Justin Fredericks: You are one of the smartest and most skilled technical people I have seen in my career, and I was lucky enough to have you as a mentor very early on. You helped teach me about Microsoft Exchange, Active Directory, and FSMO roles. You played D&D with me. You were more than a colleague; you were a friend to me. This foundational interaction early in my career helped solidify my path.

Miles Technologies: I was lucky enough to have not one, but two employment stints at this firm. It was here that I really spread my wings for the first time and knew I could succeed long term in a technology career. Some of the brightest and most skilled tech professionals I ever met were from this firm, including Ken Pyle, Chris Clancy, Ryan Armstrong, Bob Doohaluk, and Ray Gasnick

III. I also want to thank Dan Carpenter, John Politsky, and Anthony Duarte for their leadership and guidance here.

Exigent Technologies: I cannot express how grateful I am for the long tenure and experience I had at this company. The leadership at this organization influenced me tremendously, and their dedication to their customers, delivering technical excellence, and industry innovation is an inspiration. To Dan Haurey and Gerry Busardo, thank you both for every opportunity I had, every conversation I've ever had with you both I walked away having learned something – your leadership and stewardship of Exigent have made it an incredible organization, both to work for and to its customers. To Chris Jastrzebski, thank you for instilling in me the customer-service mindset I have for service delivery today. To Frank Vizzuso and Eric Burke, your commitment to technical excellence and industry thought leadership has been invaluable to my learning and career growth.

Connect Cause: In the midst of a global pandemic, you took a chance on me. Your mission to provide non-profits with exceptional service and technical guidance is profound and noble, and I am happy I had the opportunity to be a part of that team. To Paul Bender, with whom I also worked with at Exigent, thank you for your trust and investment. You pushed me to be the best professional version of myself. To Ahmmad Eied, thank you for the trust you placed in me to help steward your companies; thank you for your generosity and for the opportunities you created. Your entrepreneurial drive, business acumen, and outside-the-box thinking continue to impact me.

CPP Associates: Joining CPP Associates marked a pivotal moment in my career, allowing me to expand my expertise in managing and scaling a true managed services operation. You foster a culture that blends technical and business excellence with a "work hard, play hard" attitude. To the executive team— Pat and Paul O'Dell, and Dan Hogan—you embody a leadership style that strikes the right balance between trust, autonomy, and accountability. You've built an environment where high expectations drive success—without the burden of micromanagement. CPP has given me the incredible opportunity to lead and shape a growing Managed Services practice in an environment that blends deep technical expertise with a strong business mindset. I'm proud to be part of a team that values innovation, collaboration, and delivering meaningful solutions to clients.

And finally, to all those who are striving to build fulfilling, successful, and lucrative careers: may this book provide you with the tools, insights, and inspiration you need to achieve your goals.

I wouldn't be where I am without each of you—thank you for shaping my journey and inspiring the road ahead.

All My Best,

Aaron Cervasio

Foreword

When Aaron first told me he was writing a book, I didn't just think it was a good idea, I thought it was overdue. I've worked in and around IT long enough to know that most people either don't have the time or the guts to say what actually needs to be said about this industry. Aaron does.

The IT Factor isn't your typical "let me teach you how to be a tech guru" fluff. It's not packed with buzzwords or padded with LinkedIn clichés. It's sharp. It's honest. And it's written by someone who's had to earn every inch of credibility he's got. Not by shouting louder, but by doing the work, solving the problems, and learning the hard way when the manual didn't cut it.

Aaron came into CPP's world during a chaotic time. New department, new responsibilities, no roadmap. A lot of people would've coasted or crumbled. He didn't. He adapted fast. He asked the right questions, took feedback seriously, and figured out how to get the job done without making a show of it. That's a rare quality in any field, but especially in IT and cybersecurity, where ego often outweighs execution.

And what impressed me most wasn't just that he got up to speed quickly. It was that he didn't stop there. He didn't just want to "do the job." He wanted to understand how the whole machine worked: business operations, client expectations, human behavior, even the ugly parts—vendor politics, burnout, red tape, all of it. He's always looking around corners, seeing the downstream effect of a technical decision before others even realize there's a risk. That kind of systems-level thinking is what separates the people who survive in this industry from the ones who actually shape it.

That's what this book is really about.

It's not a guide for people who want to check boxes. It's a wake-up call for people who want to build a real career. It cuts through the noise and speaks plainly about the tradeoffs, the traps, and the moments where your character matters more than your certifications. It won't hold your hand, and it's not interested in making you feel good for just showing up. If you're looking for a shortcut, look elsewhere. But if you want to understand how to bring real value to the table—how to build trust, think critically, and get people to actually listen when you speak—then this is the book you want.

There's no gimmick here. No secret formula. Just hard-earned insight, plain talk, and a hell of a lot of lessons that too many people learn too late.

If you're early in your career, this book will help you avoid a lot of unnecessary pain. If you're mid-career, it'll give you the language for things you've felt but couldn't quite articulate. And if you've been around forever, it might just remind you why you got into this line of work in the first place. As we often say at CPP, professionals hate to be told, but they do love to be reminded!

Do yourself a favor, invest in yourself and read the book. You'll walk away smarter, and probably a little more honest with yourself about where you are relative to your peers who are shaping the industry.

Pat O'Dell
CEO, CPP Associates

Preface

Technology professionals are everywhere, but few truly stand out. In an industry where IT and cybersecurity are the backbone of almost every organization, being average isn't enough. The sheer scope of technology and cybersecurity is only outmatched by their potential to continue growing. Thirty years ago (1995 for anyone counting), most people never even heard the word "Internet." Today, billions of people walk around with it in their pockets. As pervasive as technology is, most people still aren't very technical. And for those who are, most are only passingly so.

Now, to be clear, this book isn't for casual readers looking to brush up on tech skills. It won't teach grandpa about Wi-Fi or bank fraud scams. It's not for tech enthusiasts who just want the latest gadgets, smart home upgrades, or AI tips.

You're probably thinking, **"OK, sure, but who should actually be reading this book, then?"** The short answer is "**Any technology professional or student looking for an adrenaline shot to their technology career, to upgrade and take it to the next level.**"

Take a look and see if you fall into one of these categories:

✓ **Students & graduates** seeking a strong start in IT.

✓ **IT pros** struggling to break through career plateaus.

✓ **Consultants & entrepreneurs** working to retain clients.

✓ **Job seekers** who aren't landing interviews or who choke when they do.

✓ **Experienced tech pros** who are great at their jobs but struggle to articulate their value.

✓ **Directors, VPs and managers** looking to take the lessons here and impart them to those they are leading.

If you identify with any of this, then this book is most definitely for you. If none of this applies, read it anyway. I'm confident that there are some nuggets of wisdom you can add to your professional toolbelt.

Here's the hard truth: many so-called IT professionals are barely more capable than a power user. I've seen it firsthand: tech pros who struggle to implement solutions or lack the skills to stand out. So, what separates the elite from the average? That's exactly what we'll explore in this book.

I wrote this book because I've seen talented professionals hit roadblocks, stuck in their careers, unable to earn that next promotion, or struggling to make ends meet. I've watched tech pros lose out to less capable people, not because of a lack of skill, but a lack of the right strategies. This book is here to change that. I want to share my knowledge and experience with anyone that would find it valuable… whether you are looking to enter the workforce in a tech-centered career or you're an experienced professional looking for a way to move up to that next level or who feel left behind.

There are far too many facets of technology for anyone to master them all. But, despite this, as I've developed my own career, I have learned how to become a near-indispensable resource, how to climb the ranks both in title and salary, how to be at peace with imposter syndrome, and how to provide exceptional service to those I'm working with.

So, why should you listen to me? I'm Aaron Cervasio—an IT & security professional with 20+ years of hands-on experience in tech, cybersecurity, and leadership.

I got my first computer when I was in 6th grade. At 12 years old I was the only person in my immediate family really using computers. I was fascinated with them, the Internet (in its relative infancy), and, of course, computer games. I quickly became adept at troubleshooting issues and learning about common computer problems and concepts, such as viruses, performance issues, as well as different computer components like memory, processor, and hard drives. As I went into high school, I started building my own computers (often for the purposes of playing games) and became a go-to resource for family and friends who needed help with their computers and other tech products like televisions and sound systems. I also learned about home Internet routers and very basic networking concepts.

After high school, I skipped the traditional 4-year college route and enrolled in a Networking Systems Administrator program at a technical institute (finally adding structure to my self-taught knowledge). Then, in January 2005, I started my first full-time job in information technology making $9 an hour. The job was at a very small technology company (who was outsourced to by Dell) for onsite, in-warranty computer and television repairs and tech support. Users in wealthy homes often needed large plasma television main boards replaced or computer hardware components replaced. I would also consult with them about their home networks and provide value-add consultation.

I soon realized that my skills were developing even more, and I began looking for new work that challenged me and allowed me to grow. That first 18 months I jumped around, working for a couple mom-and-pop computer repair shops and even got some very limited exposure to business networks. Then, in the summer of 2006, I started working for a Managed IT Services Provider ("MSP" for short, it's a type of technology company that provides outsourced IT and cybersecurity services for a wide variety of organizations, often industry agnostic and varying in size). It was here that I was first really introduced to other talented technology professionals and given the opportunity to learn and gain hands-on experience with servers, switches, and firewalls.

Since 2006 I have worked for a handful of MSPs. My longest tenure (so far) was almost a decade, from 2011 to 2020. At this MSP, I worked my way up from a standard remote and onsite support tech to a higher-level Systems Engineer, NOC Administrator, Technical Account Manager, Presales Consultant, and unofficially as a "vCIO" (virtual Chief Information Officer).

Unfortunately, in 2020 when the COVID pandemic hit, I was furloughed. During that time, I took the opportunity to apply my years of knowledge and experience to obtaining certifications while searching for a new job. I was able to obtain multiple technical and cybersecurity certifications, including the

Certified Information Security Manager (CISM) credential. I was quickly hired by another Managed Services Provider leading the Technical Services department. In that role, I was responsible for building out processes and procedures, writing IT and infosec policies for both our organization and our clients, and developing our internal suite of products and services. I also provided top-level escalation support for projects and service tickets, while serving as both a vCIO and vCISO (Virtual Chief Information Security Officer). In those capacities, I consulted with clients on their technology, IT operations, and information security concerns—particularly those tied to regulatory compliance requirements.

After three years, I was promoted to Chief Information Security Officer. In this role, I researched and developed our internal infosec programs and led the creation of a new, standalone information security division. This included launching dedicated products and services for our customers and prospects, all in support of significant business development goals (and with the broader mission of making the world, our businesses, and our lives just a little bit safer).

Since then, I've moved on to build out and lead an entire Managed Services division for a mid-sized tech company. In this role, my responsibilities included choosing key vendors and partners to build out a comprehensive and diverse portfolio of IT and cybersecurity solutions, executing on their implementation, and leading the teams to ensure top-notch service delivery. It's all become substantially more lucrative than $9 an hour.

I hope that sharing my backstory here earns me some credibility and instills in you a sense of confidence as you delve deeper into the book. I plan on sharing the collection of this experience and knowledge, hard-earned over a long career, by dispensing a wealth of advice and guidance through simple, modern-day fables with relatable characters and injecting some author commentary (among other fun things and banter) along the way. It's my hope that as you read these fables and comments, you will identify with various characters and their situations, that you will see common threads between these stories and your own, and that you will find my personal input about these stories valuable for your own development.

Here's the journey we're about to embark on and how the stories we'll explore will illuminate the path to your success (in no particular order):

- We'll journey through the process of crafting a compelling resume, acing interviews, negotiating salaries, and setting career goals. Through practical examples and solutions, you'll be equipped to overcome common career challenges.

- We'll delve into the realm of essential soft skills, such as communication, teamwork, adaptability, scalability, and problem-solving. Through engaging narratives, you'll discover practical ways to hone these skills and understand their impact on your professional growth.

- We'll confront the notorious work-life balance conflicts in IT and cybersecurity head-on. We'll explore strategies to maintain a healthy balance, mitigate burnout, and uphold high standards of accountability.

- We'll navigate the importance of continuous learning, keeping abreast of industry trends, and investing in ongoing education, certifications, and skill development. Our stories will underscore the value of staying ahead of the curve in the dynamic IT landscape.

- Through tales of connection and collaboration, we'll highlight the power of networking and building relationships within the industry. You'll gain insights on effective networking strategies, both in the digital world and beyond.

- We'll address imposter syndrome and how to maintain a healthy mix of confidence and humility, how to know your worth, and how to identify if and when you're being taken advantage of.

Lastly, I want you to know I've been where you are today. Throughout my career, I've faced numerous challenges and setbacks, but with determination and a passion for learning, I overcame them and advanced in the IT and infosec industries. I'll be using a combination of these story-telling narratives and addressing you directly to dispense guidance and advice. These fables, though embellished or silly for narrative effect, will be rooted in my direct experience, and offer valuable lessons and practical input to help you overcome your own career obstacles.

This book gives you the mindset, skills, and strategies to rise above the noise and build a career that commands respect, promotions, and opportunities. The knowledge is here. Your success starts now. Let's get to work.

Table of Contents

Part 1

Opening a Door and Stepping Through

Premise

The network was down. Again.

Inside Lee Health Solutions, the staff (nurses, physicians, and office administrators) were growing restless. The patient management system was frozen, and frustrated front desk workers were scrambling to check people in manually.

Dr. Lee stood by the reception desk, arms crossed, watching the chaos unfold. He knew this wasn't the first time, and at this rate, it wouldn't be the last.

In the back office, the IT team was scrambling to figure out what had gone wrong. Someone suggested rebooting the firewall. Another ran a diagnostic, only to be met with more error messages. No clear answers. No leadership. No direction.

It was just another day of tech struggles at one of the few remaining independent medical practices in Keystone. This company was a business that had managed to survive in an era where corporate healthcare networks consumed smaller offices like hungry sharks.

But Dr. Lee knew the practice couldn't survive without technology. And at that moment, technology was failing them. Something had to change.

Lee Health Solutions was a private medical practice in the heart of the bustling city of Keystone, nestled among towering skyscrapers and crowded streets. It was a well-respected institution, known for its excellent patient care and innovative medical solutions. Unlike most private practices in the area, Lee Health Solutions has remained one of the few not gobbled up by a larger, corporate health network. Untarnished by large-scale enterprise bureaucracy, it was a small fish in a large pond, but appealing in its own way for patients and employees alike. However, behind the scenes, the technical team that supported the practice was struggling.

The team was a mix of talented individuals, each with their own unique skills and experiences. But despite their individual capabilities, they were failing to function as a cohesive unit. Communication was poor, morale was low, and there was a distinct lack of direction. There were also inter-departmental challenges that exacerbated the issues. The team was in desperate need of change.

Dr. Lee, the founder of the practice, watched all of this unfold with growing concern. He understood the critical role that the technical and security team played in the success of the practice. He knew that without a functional tech team, the practice could not provide the level of care that their patients deserved. He decided it was time to act.

Dr. Lee knew that the team needed a leader, someone who could guide them, inspire them, and help them reach their potential. He decided to hire a Director of Technology and Security, someone with the skills and experience to turn the struggling department around.

The task at hand was not an easy one. The technical department at Lee Health Solutions was not just responsible for the day-to-day IT operations, but also for ensuring the practice's compliance with stringent HIPAA regulations and maintaining robust information security controls. The stakes were high, and the challenges were many. But Dr. Lee was determined to find the right person for the job.

And so, the search for a new Director of Technology began. Little did Dr. Lee know the perfect candidate was just around the corner. His name was Scott.

Technology is the backbone of nearly every modern business, and without a strong IT and security team, even the best companies can struggle.

I chose a private medical practice in a fictional city for this example because real-world business challenges go far beyond technical problems. Not every company is a well-oiled machine, and understanding the context of an organization's struggles can make you a more valuable candidate.

When looking for a job, it's not just about finding an open role. It's about understanding why that role exists in the first place. Is the company expanding rapidly and in need of better infrastructure? Is it struggling with inefficiencies or compliance issues? These factors can tell you what kind of challenges you'll be stepping into and whether your skills are the right fit.

For example, if you're applying for a position in healthcare IT, having a strong grasp of HIPAA compliance, patient privacy concerns, and security best practices will make you far more appealing than someone who only lists generic IT skills. Employers want to see that you understand their world and can help them solve the specific problems they face.

Before applying for any job, ask yourself...
- Is this company growing or struggling?
- How do my skills fit their current challenges?
- What unique value can I bring that sets me apart from other candidates?

Being intentional about these answers will make you a stronger, more competitive applicant. It will also increase the chances of landing a role that aligns with your expertise and career goals.

The Job Search and Hiring Process

Dr. Lee had seen his fair share of resumes over the years. He knew what to look for, and more importantly, what to avoid. As he sifted through a pile of applications for the Director of Technology and Security position, he found himself shaking his head more often than not.

Some resumes were filled with jargon and buzzwords but lacked substance. Others were poorly formatted, with typos and grammatical errors that made them difficult to read. A few candidates had impressive technical skills but lacked the soft skills necessary for a leadership role. Some resumes were 5 pages long and difficult to read through, while others were condensed to half a page and left more questions than answers. It was clear that many of the applicants had not taken the time to tailor their resumes to the position and possibly didn't even know the industry they were applying to.

Taking a gamble, Dr. Lee called a few of the applicants anyway. Despite being skeptical about the quality of their resumes, Dr. Lee was desperate to fill the role and was hopeful there might have been a standout interviewee. Unfortunately, the interviews were no better than the resumes themselves.

One candidate joined a remote interview 10 minutes late and was driving in his car the entire time. The background noise made it nearly impossible to hear him. First impression? Immediate disqualification.

Another walked into the in-person interview wearing jeans, sneakers, and a polo. Not the end of the world, but when he dismissively ignored the front desk secretary, Dr. Lee knew he wasn't leadership material.

A third emailed an hour before the interview to reschedule. Emergencies happen, but this last-minute request showed poor planning and disregard for others' time.

A fourth simply couldn't answer basic technical and leadership questions, proving his resume had been more buzzwords than substance.

Dr. Lee sighed, rubbing his temples as he stared at the dwindling pile of resumes. Weeks of searching, endless interviews, and still no clear answer. He knew that hiring for technical roles was never easy, especially when leadership skills were just as important as IT expertise.

"Maybe I'm being too picky," he thought. But deep down, he knew settling for the wrong candidate would only make things worse.

Just as Dr. Lee was beginning to lose hope, he finally came across a resume that stood out from the rest. It was from a candidate named Scott. His resume was clean and well-organized, with a clear career trajectory and a balance of technical and soft skills. He had a strong background in IT, with experience in both regulatory compliance and cybersecurity. But what really caught Dr. Lee's attention was Scott's emphasis on continuous learning, technical and procedural writing, and team development. It was 2 pages long, detailed enough to show a long career and appropriate skills while trimming the unnecessary excess. It also included a very brief cover letter with Lee Health Solutions logos on the letterhead.

Intrigued, Dr. Lee decided to invite Scott for an interview. He was eager to see if Scott was as impressive in person as he was on paper.

It wasn't long before the day of the interview arrived. Scott walked into the Lee Health Solutions office with a sense of calm confidence. He was dressed professionally, but not overly formal—a blue button-down shirt, no tie, black slacks, a black jacket, and a nice pair of black shoes. Scott struck the right balance for a tech leadership role.

Dr. Lee started the interview with some standard questions about Scott's background and experience. "Can you tell me about a time when you faced a significant challenge in your previous role and how you handled it?" Dr. Lee asked.

Scott paused for a moment, collecting his thoughts carefully before answering. "In my last role, we had a major security breach. It was a stressful time, but I learned a lot from it. I led the team in identifying the source of the breach, resolving the issue, and then implementing new technical and operational measures to prevent it from happening again. It was a team effort, and I was proud of how we handled it. Thankfully the data was all backed up securely, but it did take a lot longer to find and recover than we would have liked. I wound up overhauling the backup and recovery procedures and configuration after that to address the shortcoming and now use that as a benchmark for my recommended standards."

As the interview progressed, Dr. Lee was impressed by Scott's depth of knowledge and his clear communication skills. But what really stood out was Scott's understanding of the unique challenges of IT in the healthcare industry.

"You've mentioned HIPAA compliance a few times," Dr. Lee noted. "Can you talk a bit more about why that's important in your role?"

"Absolutely," Scott replied. "In healthcare, we're dealing with sensitive patient data. It's our responsibility to protect that data and ensure we're in compliance with all regulations. It's not just about avoiding fines. It's about maintaining our patients' trust."

Scott quickly earned himself a second interview, which happened a few days later. Towards the end of the second interview, the topic of salary and benefits came up. Scott was well-prepared for this discussion. He had done his research and knew the market rate for his role.

"I've done some research," Scott said confidently, making sure to maintain eye contact. "Given my experience in cybersecurity, HIPAA compliance, and IT leadership, I believe a fair range for this role is $120,000 to $140,000. I know this is slightly above the posted range, but I'd love to discuss how we can align compensation with the responsibilities of this position."

Dr. Lee appreciated Scott's transparency and professionalism during this often-difficult conversation. Many candidates fumbled through this discussion. Scott, however, had come prepared.

By the end of the interview, Dr. Lee was convinced. Scott was the leader that the IT team needed. He offered Scott the position, and after some negotiation, they agreed on a competitive salary and benefits package. Scott was officially the new Director of Technology and Security at Lee Health Solutions.

Finding the right role, crafting a strong resume, and negotiating a fair salary are all critical to advancing your career. This isn't just true for IT. It applies to every industry. But in IT and cybersecurity, the competition is fierce, and small mistakes can cost you opportunities.

A bad resume can get tossed aside in seconds, while a strong, well-structured one can open doors.

What NOT to Do in a Resume...

- ✗ Too long. A 5+ page resume overwhelms hiring managers.
- ✗ Too short. Half a page with barely any detail leaves questions.
- ✗ Buzzword-heavy with no substance. If your resume is just a list of acronyms, you're doing it wrong.
- ✗ Poor formatting & typos. Spelling and grammar mistakes make you look sloppy.

What WILL Make Your Resume Stand Out...

- ✓ 1-2 pages max. Enough detail to showcase experience without fluff.
- ✓ Tailored to the role. Highlight skills that match the job description.
- ✓ Easy to scan. Clear sections, bullet points, and logical structure.
- ✓ Personal touch. A customized cover letter (like Scott's letterhead logo) can set you apart.

A well-crafted resume isn't just about listing experience. It's about storytelling. I once saw a resume styled to look like an Amazon shopping page, complete with:

➢ 5-star reviews (references).
➢ "Add to Cart" (email to hire the candidate).
➢ A product description listing their IT skills & achievements.

Now, you don't need to be a photoshop expert and go to that level of uniqueness. But the concept, thinking outside the box, and doing something to ensure your resume stands out against the competition – that's what you're going for. Just like how Scott chose to write a cover letter and create a unique letter head, this type of personal touch can pay dividends back when being considered for a job role.

Once your resume is attention-grabbing, you'll get interviews. Don't blow them.

What NOT to Do in an Interview…

✖ Show up late. First impressions matter.
✖ Multitask or be distracted. One candidate took a remote interview while driving.

- ✗ Dress too casually. A polo and jeans might be fine for engineering but not for leadership.
- ✗ Be rude to staff. If you ignore or dismiss the front desk staff, it gets noticed.
- ✗ Cancel last-minute. Emergencies happen, but poor planning raises red flags.
- ✗ Be unprepared. If you can't answer basic technical or leadership questions, you won't get hired.

What WILL Get You Hired...

- ✓ Confidence (not arrogance). Know your worth but stay humble.
- ✓ Clear communication. Explain technical concepts concisely (don't ramble).
- ✓ Engagement & professionalism. Look the part, act the part.
- ✓ Resourcefulness. If you don't know an answer, show how you'd find it.

💡 **Pro Tip**

If you get stumped by a technical question, don't guess. Admit it and explain your approach. And after the interview? Actually go research the answer and email the hiring manager. That follow-up alone can make you stand out.

When you do well enough on your first interview, you will likely go through additional rounds before any sort of offer occurs (usually). The interview is also your opportunity to prove value beyond the experience and knowledge in your resume. Recruiters and hiring managers want to know what you personally bring to the table that makes you the best person for a role you're interviewing for. Anyone can be technical or have a certification (and listen, these things are important). But certifications and project management experience alone aren't what will get you that better-paying job. It's your personality, how articulate you are, how confident you carry yourself, how humble you're capable of being, and how resourceful you are when you don't know the answer; these are what will get you hired and what will allow you to negotiate a better salary for yourself.

Embrace opportunities to take responsibility for mistakes or address shortcomings in your experience when you are interviewing. Don't do this in a way that makes you look incompetent but use these as ways to prove you learned a lesson and how that changed you, how that informs the way you would handle a similar situation in the future. For example, Scott tackled the question about a challenge he faced head-on, he spoke in specifics not generalities (e.g., this specific breach happened vs. something like 'Oh, I tend to be a perfectionist, so I spend a lot of time working after hours' which really isn't a specific challenge and doesn't tell the hiring manager anything useful). The breach that Scott discusses however is a sticky topic in a world where breaches are serious and can shake the confidence of a company or hiring manager. But he overcame that incident, actually used it to change his approach for backups, and that experience makes him even more valuable for Lee Health Solutions.

Another hurdle is that a lot of candidates struggle with salary discussions. But negotiating confidently can mean thousands more per year.

How to Negotiate with Confidence...
- ✓ Research market salaries – Use Glassdoor, Salary.com, and industry benchmarks. You can ask AI platforms as well (but always fact check them).
- ✓ Know your value – Tie your request to your skills, experience, and impact.
- ✓ Be flexible, but don't undersell yourself – Companies expect negotiations.

💡 Example Script for Salary Talks

"Based on my research of the market and my experience in A, B, and C, I believe a fair range for this role is $X to $Y. I know this is slightly above the posted range, but I'd love to discuss how we can align compensation with the responsibilities of this position."

Ultimately, you need to have done your research. Also keep in mind the field and type of organization you're working for. A relatively small non-profit organization, for example, will likely not pay as much for an IT Manager as a large enterprise corporation will (and that doesn't mean it isn't worth working at the non-profit, your experience, knowledge, and personal fulfillment may be better suited to the Suicide Prevention Group than as a faceless number at ABC Retail Giant).

Be willing to ask for what you think you and the role are worth. Be honest and confident, but not arrogant. It's a fine line to walk. But you need to be able to back up everything you say and the confidence you project with **true knowledge**. Swagger without the knowledge to back it up will end poorly. It may seem strange, but the right amount of humility and resourcefulness actually shows even more confidence – the ability to admit what you don't know and your methods for finding it.

Also keep in mind that when it comes to job hunting, data is everything. The more information you have—about the company, the role, and your value—the stronger your position.

As Sherlock Holmes famously said…

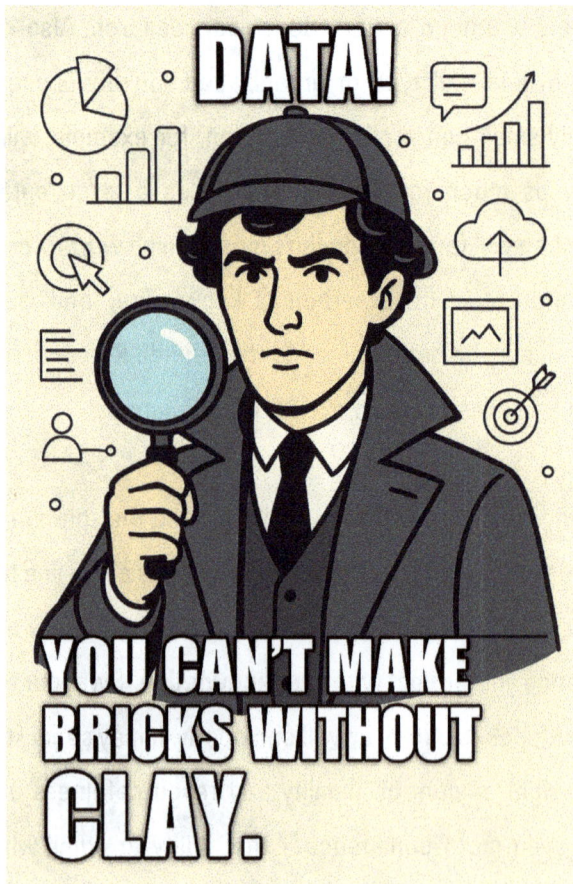

DATA! YOU CAN'T MAKE BRICKS WITHOUT CLAY.

Lastly, it's OK to think outside the box. Remember, businesses are taking a gamble on you as much as you're taking a gamble on them. But, if salary becomes a roadblock, consider this:

➢ Offer to start at a slightly lower salary on the condition that after 90-120 days, if you meet performance expectations, your pay is adjusted.
  - o This proves your confidence in your ability to deliver results and reduces the employer's risk.
  - o Many companies will respect this bold approach and agree to the terms. Make sure it's in your offer letter.

💡 **The best candidates don't just apply for jobs. They position themselves as the missing puzzle piece a company needs.**

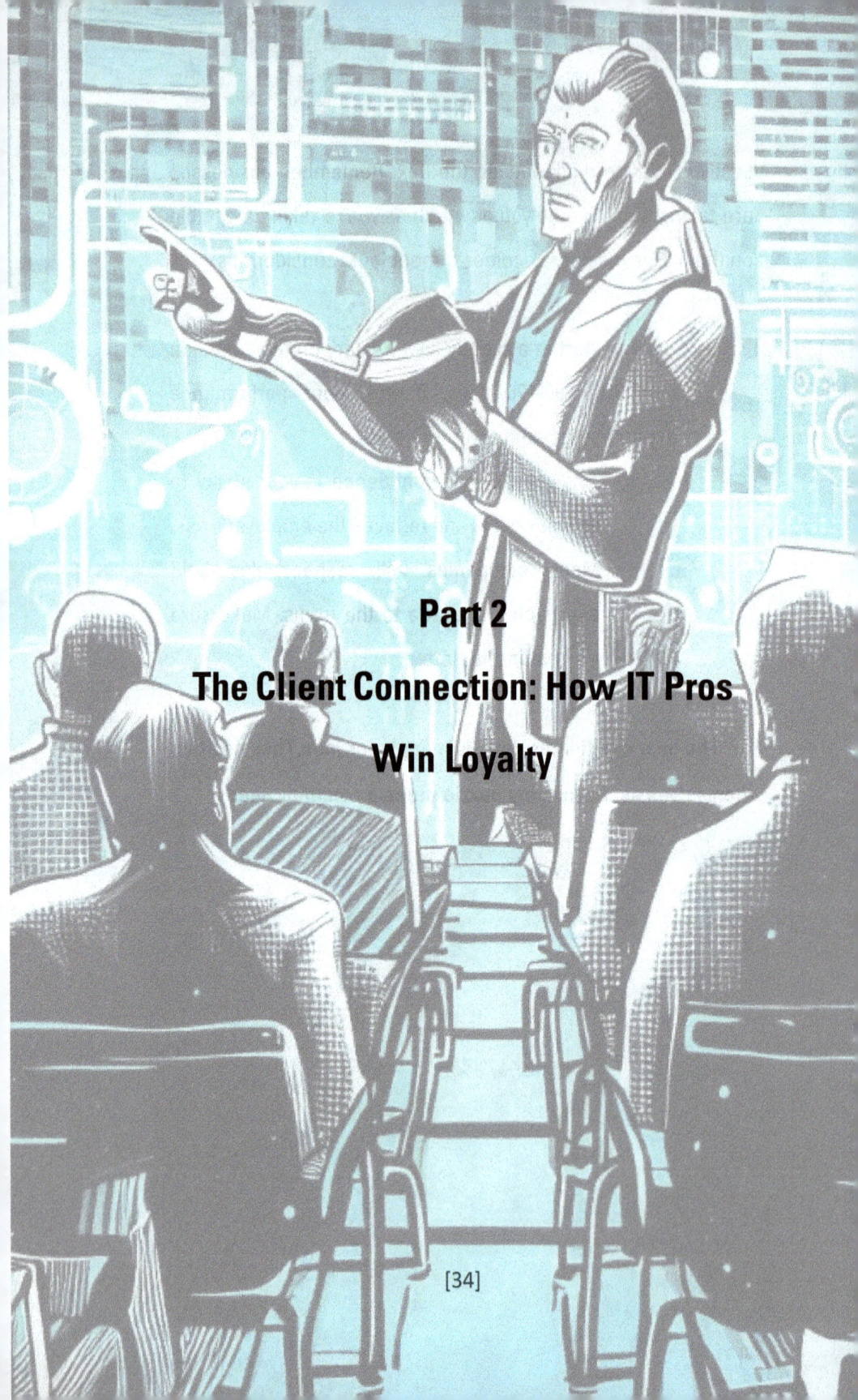

Part 2

The Client Connection: How IT Pros

Win Loyalty

Premise

In the sprawling metropolis of Los Diablos, there was a highly skilled IT entrepreneur named Tyrell who owned and operated his own firm called Eagle Eye Consulting. Before opening his business, Tyrell was well-known in the IT industry for his technical acumen and his ability to troubleshoot any IT issue that came his way. After a bout of recent layoffs due to a slowing economy, however, Tyrell found himself without a job. He decided to take the plunge and start his own managed services firm.

Not long after opening shop, a law firm named "Justice & Associates" called Eagle Eye Consulting for help.

We previously went over Resumes and Interviews. After losing a job, you may find yourself searching for a new role somewhere. But many people try to start their own consulting firms as well. I do not intend on getting into how to create and run a managed services company (perhaps I'll save that for another book). The premise of this story is much shorter because we're not going to explore job searching, resume building, or interviewing, nor are we exploring the intricacies of starting and running a business. One of Tyrell's first major projects came when a local law firm, Justice & Associates, reached out for IT help. He had the skills to solve their problem. But would his approach win them over? In IT, technical expertise is expected, but mastering client relationships is what separates the best from the rest. That's what we'll focus on in this story.

Throughout your career, individual experiences in different roles, whether freelancing or working for others (or even owning your own consultancy) will shape you. Look for the valuable lessons that failure brings in order to grow and adapt.

A Phone Call and Project

"Yeah, hello?" Tyrell saw an unfamiliar number on his phone screen but answered anyway. After all, this was the number listed on his company website.

"Ummm, sorry, I am looking for Eagle Eye Consulting? My name is—" Unfortunately, the caller was cut off by Tyrell, "OK, yeah that's me, how can I help you?"

The caller took a breath and then continued, "Yes, as I was trying to say, my name is Jessica Justice, and I am one of the partners here at Justice and Associates. We're a Law Firm and—" Tyrell cut off Jessica again. "Sorry, I don't need any legal advice right now. If you need help with a tech problem, I'm your guy, though."

Jessica, getting a bit more frustrated, continued, "Please just give me a minute to finish, I am not trying to sell you any services. We do need IT help. We've been experiencing performance issues and downtime with our server, it's a few years old, and the company we used to work with is out of business. Is this something you can help with?"

Tyrell replied, "Uh-huh. Yea, I can spin up a new server no problem. I'll send you an estimate, and once you approve it, I'll order the parts I need and put it together, then we'll migrate your old server to the new one. Not an issue."

"Oh, that's great news. Yes, please send the estimate over. How long do you think the whole thing will take?" Jessica was feeling relieved for the first time in a few weeks.

"Well, I need to check out the existing server first and get some more information, but honestly? Most things like this are easy peasy lemon squeezy. So, you're probably looking at a week or two." Tyrell confidently responded.

Tyrell had a lot of experience migrating various types of servers over the years. His confidence in the project put Jessica's mind at ease. The two exchanged contact information. Not long after, Tyrell coordinated an onsite discovery of the server environment, and found the hardware was older, out of warranty, and was running some older software that he would need to work with the software vendor to migrate. He threw together an estimate for hardware and labor costs and sent them over to Jessica. After a few days, Jessica approved the estimate and Tyrell started his work.

Unfortunately, as he delved into the project, Tyrell encountered several issues.

First, the server hardware he wanted was out of stock at all major distributors. So, it was 2 weeks before he even got the hardware he needed. Jessica had to call him after 10 days to get an update.

After the new server was finally up and running, Tyrell started the process of migrating data. He knew the old server had some backup software running on it based on his initial discovery. However, during the file migration process a critical data corruption issue surfaced, and the backups he found during the discovery had stopped working. This led to having to resort to restore data from older backups.

Then, Tyrell discovered unexpected compatibility problems with some of the firm's older software applications. He had to call the vendor for software migration support.

Each of these issues added more time to the migration process.

Meanwhile, Jessica's frustration turned into full-blown resentment. Every delay cost her firm money, and every unanswered email felt like another broken promise. She had trusted Tyrell's confidence. Now, four weeks later, her firm's productivity was suffering, and she was chasing him down for updates. Clients were frustrated, cases were delayed, and every time she called, she got Tyrell's voicemail instead of answers. The firm was left in the dark and began to question Tyrell's abilities and reliability.

When Tyrell finally finished everything and called the firm to tell them that the server migration was complete, they were not pleased. Despite the fact that Tyrell had successfully migrated their server, the lack of communication and the various delays had tarnished their experience. They paid for the migration but decided not to return to Eagle Eye Consulting in the future.

After the law firm's departure, Jessica sent Tyrell a lengthy email about their dissatisfaction and why they would not be doing business again in the future. Tyrell reflected on the situation. He realized that while he had been focused on the technical aspects of the server migration, he had neglected an essential aspect of his service - customer service and communication with his client. Despite performing a discovery and knowing the potential risks, Tyrell gave Justice and Associates a very optimistic timeline. He never told them about potential complications, and he never provided proactive updates about the progress of the migration.

Tyrell knew he needed to make changes. He started by never interrupting his clients or prospects on the phone again. He also got a separate business phone number so he would always know when to answer more professionally. When performing discoveries and generating estimates, he began including statements of work that set clear expectations with his clients about the project process, including potential issues and delays that could arise, leading to more conservative estimates. He also made it a point to provide regular updates to his clients via milestone meetings, keeping them informed about the progress of their projects.

Over time, Tyrell noticed a significant improvement in his business. His clients appreciated his transparency and regular communication. They felt more confident in Eagle Eye Consulting's services and were more understanding when issues arose. Tyrell's consultancy flourished, and he learned a valuable lesson in professional growth.

When people think of IT professionals, they often picture individuals hunched over keyboards, deeply engrossed in lines of code or some complex system configurations, working late into the night. And many times, we are that. However, while these technical aspects of the IT world are certainly important, the true heart of IT success lies in customer service. It all starts with treating every user you interact with as if they were a personal client of yours.

Don't view technology or your technical skills as the value you're providing. Instead, think of yourself as a customer service representative, and the skill you're providing is excellent customer service. The product or service you're providing customer service for only happens to be technical in nature. Providing exceptional customer service is the foundation upon which long-term satisfaction and success are built. You must hone your ability to manage expectations, communicate effectively, empathize with clients, and think about scalable solutions, ultimately bringing higher levels of satisfaction for your clients and organization, and leading to your long-term success.

Misaligned Expectations Can Be Disastrous.

One of the most challenging aspects of a technical role is managing expectations. Clients often have high hopes for quick fixes and seamless solutions, but as we know, technology doesn't always cooperate. The keys to managing expectations are to be transparent, proactive, and consistent in your communication.

In the story, Tyrell downplayed the potential challenges he could face. He had not adequately managed the client's expectations from the beginning. They became increasingly frustrated with each delay. In their eyes, Tyrell was not meeting the deadlines he had promised, and their trust in his ability to deliver began to wane. They questioned whether he even architected the project properly to begin with.

In hindsight, Tyrell realized he should have been more thoughtful about unforeseen complications (whatever they may be) and more transparent about the fact that unknown challenges were something he might have faced during the migration. He could have then provided a more conservative estimate of the timeline. Had he communicated more proactively, the client would have been better prepared for the possibility of delays, and their frustration could have been mitigated. By doing this,

Tyrell would have created a strong foundation of trust and understanding, even when things didn't go according to plan.

Providing regular status updates and maintaining open lines of communication throughout any project can help to further reinforce trust and keep your clients informed and engaged. Transparency is crucial to building trust with your clients and colleagues. From the outset, clearly explain the scope of work, the steps you'll take to address the issue or project, and a realistic timeline for completion. Be honest about potential challenges and limitations and avoid over-promising or setting unrealistic expectations.

Proactivity is also essential in demonstrating your commitment to resolving an issue or executing a project. Take the initiative to engage with the client, keep them informed, and address any concerns they may have. Regularly reassess the situation and be prepared to adjust if necessary.

Consistency in your communication ensures that clients feel heard, valued, and confident in your ability to deliver the desired results. Provide updates on the progress, taking care not to let too much time pass, to keep your client in the loop and reassure them that their issue is being actively addressed.

If complications arise, promptly inform your client and discuss any adjustments to the plans you put in place. This further demonstrates your commitment to transparency and helps to build trust with your clients.

By being transparent, proactive, and consistent in your communication, you'll effectively manage client expectations, reduce the likelihood of disappointment and frustration, and foster a more positive customer experience.

If Tyrell had done all of this, he may have come out with a new client and more business, rather than a dissatisfied prospect that he forever lost future business with. The technical result of the project would have been the same in either case – the difference maker here was not in the technical ability to perform the work, but all of the soft skills surrounding the technical work itself.

If you want to stand out as an exceptional resource wherever you work, you need to be more than just a technical savant.

(Moving up in the context of a technical career is not just about how well you can make a computer go 'beep boop beep' but rooted in thinking about service delivery to any end user or client from a customer service lens.)

IT isn't just about fixing problems, it's about winning trust. The best tech professionals aren't just problem-solvers; they're trusted advisors. If you want to level up your career, don't just be good at the technical side. Be the person clients turn to in a crisis. Be the person they recommend without hesitation. Because in the end, one of the most valuable skills in IT isn't just knowing how to fix things—it's knowing how to build relationships.

💡 Key Takeaways from this Chapter

- ✓ Your technical skills are expected, but client relationships will set you apart.
- ✓ Set realistic expectations from the start; be honest about risks and timelines.
- ✓ Communicate proactively; regular updates prevent frustration and build trust.
- ✓ A delayed project isn't always a failure, but losing client trust is.
- ✓ The best IT pros aren't just problem-solvers. They're trusted advisors.

Part 3

The Art of Making IT Simple

Premise

In the city of Sherwood, there was a regional bank named Lionheart Financial. Six months ago, they hired a new help desk support technician named Robyn. Robyn was a quick learner and had a knack for solving IT issues. However, her true strength lay in her exceptional communication skills and her ability to simplify complex concepts for her clients.

Robyn's skills didn't go unnoticed. Within those six months, she was promoted to Team Leader, a position that many thought would go to Phil, a more senior employee who had been with Lionheart Financial for several years. Phil was known for his technical expertise, but he tended to be a bit gruff in his communication and often struggled to explain complex issues in a way that clients could understand. Everyone knew Phil was thorough and detailed, but they could never make sense of what he did to solve their problems.

One day not long after Robyn's promotion, Phil approached the regional manager and his boss, Richard, his frustration evident. "Rich, I don't understand," he said, his voice tinged with bitterness. "I've been here for years, and I've solved more technical issues than I can count. I think Robyn's a great asset for the company, but I feel like I should have gotten that promotion. Why did you promote her instead of me?"

Richard looked at Phil, taking in his frustration. He could have gotten defensive, but he chose to listen, understanding that Phil's frustration came from a place of confusion and disappointment. "Phil," he said gently, "I think it's not just about the number of issues you solve or your tenure. It's about how you solve them and how you communicate with our users."

Phil frowned, not quite understanding. So, Richard decided to share some feedback with Richard directly as well as a story about a user who Robyn had helped, which ultimately led to her promotion.

We previously discussed setting expectations and customer service. Those critically important aspects of career success are built on a foundation of exceptional communication skills.

Whether written or spoken, the way you communicate will impact the perception of others and this often directly translates into how people feel about those interactive experiences. For better or for worse, career success is not just measured in how well you do a job technically, but also how you make the people around you feel while doing it. If you take two equally-as-good technical people and compare them, the difference maker will be in how you make others feel when your work impacts them. Heck, in many cases you could have two technicians where one is better technically than the other, but the first person still makes people feel better after their interactions. The less technical person will still generally be favored by leadership and users alike, the result being that they will move up more quickly.

This story will explore how Robyn was able to make the end users feel, leading to unsolicited praise and her promotion while Richard is still in the same old role.

Complexity Doesn't Need to Be Complicated

Phil took a seat in Richard's office. "Phil, last week, we got a complaint from Marian that she was down for several hours, and she had to work late to finish out her reports." Richard sat down as well behind his desk while speaking.

"What was the complaint? I told that entire department that the information corruption needed to be repaired, which required us to take it offline so we could utilize a proprietary diagnostic tool from Cyberdyne Systems in order to find and eliminate any improperly truncated entries in the SQL database!" Phil was clearly getting angry, as his voice was raised.

Richard waited patiently for Phil to finish before responding. "Phil, this is part of the problem. I have no idea what you just said. It's gibberish to me. And it was to Marian as well. We all get it, you're a technical genius. And no one is questioning the technical work you do. But at the end of the day, Marian was still left having to work late and still didn't understand what the actual problem was."

There was a brief silence before Richard continued. "Let me tell you about John's experience with Robyn."

John had been experiencing a persistent issue with his computer that he described as his "Internet crashing multiple times a day." Several technicians, including Phil, had tried to help John through email, providing him with technical explanations and solutions, but the problem persisted.

Recognizing the need for a different approach other than email, Robyn decided to call John instead. She knew that John was frustrated, but she also knew that speaking directly to a live person would make John feel more supported. After letting John vent his frustrations, Robyn assured him that she understood his problem and would make resolving it her top priority.

Robyn then began asking John some probing questions to better understand his issue. Instead of relying solely on technical jargon, Robyn used simple language and real-life examples to help John understand the problem. For instance, she asked, "When your Internet crashes, can you still work on other programs like your spreadsheets or word documents?"

John's response revealed a crucial detail that the previous technicians had missed: his screen would flash blue, and his computer would reboot every time his "Internet crashed." With this information, Robyn suspected that the issue was not with John's Internet, but with his computer hardware.

To confirm her suspicion, Robyn used WinDBG (a Windows Debugger tool) to analyze the memory dumps from John's computer. The analysis confirmed that the crashes were due to faulty memory. Robyn explained this to John using an analogy, saying, "It's like your computer has a diary where it writes down why it's having a bad day. I just read the diary, and it seems like your computer is struggling with its memory."

John appreciated Robyn's simple explanation and was relieved to finally understand the root cause of his issue. Robyn helped John get a new PC, and throughout the process, she remained patient, persistent, and respectful of his needs and preferences.

After telling the story, Richard continued, "After the issue was resolved, John sent an email to me expressing his gratitude, Phil. He appreciated Robyn's communication style and her effort to understand and resolve the issue."

"So, Phil," Richard concluded, "it's not just about how long you've been here or how many issues you've solved. It's about how you solve them and how you communicate with the users. That's why Robyn was promoted."

Phil was silent for a moment, then nodded. "I see," he said. "I guess I have some things to learn from her." From that day forward, Phil started paying more attention to how he communicated with users, and he saw a noticeable improvement in his relationships with them. He realized that Robyn's promotion was not a slight against him, but a testament to her skills and the importance of effective communication in their line of work. It wasn't long before Phil himself was promoted as well.

John's experience with Robyn in our story reinforces the importance of mastering simplification and diversified communication as essential techniques for IT success. It also demonstrated that being a great tech goes beyond technical expertise. Effective communication is key to delivering exceptional customer service and building strong relationships with clients.

When communicating with peers and end users alike, it's important to be patient. Take the time to explain your thought process, the steps you're taking, and the reasoning behind your decisions. Encourage questions and be open to feedback.

Another important part of good communication to remember is that, as an IT professional, you're fluent in a language that many of your clients don't speak. To ensure their satisfaction and understanding, it's essential to bridge the gap between technical jargon and everyday language.

This means breaking down complex concepts into simpler terms, using analogies or real-life examples to make the information more accessible.

During my time as a help desk technician, I recall that many users would commonly describe issues in a way that truly didn't capture what was happening. Asking probing questions is an invaluable skill that helps you better understand your clients' pain points and get to the heart of their problems. By delving deeper into the issues they're experiencing, you're better equipped to identify the root cause and develop appropriate solutions. Alongside this, active listening plays a crucial role in uncovering additional issues that may not have been originally reported. As you attentively listen to your end users or clients, you may discover hidden concerns, unvoiced expectations, or underlying frustrations that can be addressed to improve their overall satisfaction.

Incorporating probing questions and active listening into your communication repertoire not only enhances your problem-solving abilities but also demonstrates your commitment to fully understanding and addressing your clients' needs. By doing so, you will strengthen your relationships with clients and elevate your status as an indispensable IT professional.

It's also supremely important to use multiple contact methods when trying to get in touch with a client or user. This ensures that your message reaches them through their preferred channel and demonstrates your commitment to keeping them informed. For example, you may call and leave a voicemail if they do not answer, but also follow up the voicemail you leave with an email. This approach increases the likelihood that they will receive your message and respond in a timely manner. Conversely, don't just rely on email to follow up, especially if you have not gotten a response. In such cases, consider reaching out through a phone call, instant messaging, or even a text message, depending on the relationship and the urgency of the matter.

By diversifying your communication methods, you show clients that you're committed, adaptable, and attentive to their needs. This approach not only helps to manage expectations but also contributes to building trust and fostering a strong, collaborative relationship. Just be sure to respect your client's boundaries and adhere to any communication guidelines they may have in place. Ultimately, effective communication is about being flexible, persistent, and respectful of your clients' needs and preferences.

💡 **Key Takeaways from this Chapter**

✓ Technical skill alone won't get you promoted. Communication matters just as much.

- ✓ Avoid jargon! Use analogies and simple explanations to bridge the gap for non-technical users.
- ✓ Active listening and asking probing questions are crucial for solving real problems.
- ✓ Diversify communication. Use emails, calls, and instant messaging to ensure clarity, efficiency, and resolution.
- ✓ The best IT pros don't just fix things, they make people feel confident and supported.

Part 4

The Power of Empathy in IT: Why

Clients Remember YOU

Premise

In the city of Gillikin, a vibrant hub of commerce and industry, there was a logistics company named Yellow Brick Transportation. This company was not terribly large (about 100 employees), but it played a crucial role in the local economy. Yellow Brick Transportation was responsible for coordinating and managing the transportation of goods for numerous businesses in the region, ensuring that products reached their destinations on time and in perfect condition.

One of the key employees at Yellow Brick Transportation was a woman named Dorothy. Dorothy was a logistics coordinator, and her role was to manage and oversee the transportation orders that flowed through the company. Every day, she would work with a specialized software application called "Emerald" to schedule, track, and manage these orders. The Emerald software was a comprehensive tool that allowed Dorothy to manage all aspects of the transportation process, from scheduling pickups and deliveries to tracking shipments in real-time and managing the company's fleet of vehicles.

One day, Dorothy was working on an important order for a large customer, Flying Monkeys Studio, located in Quadling County. This order was particularly important as it involved a large shipment of unique equipment that was needed for a major production. However, in a moment of distraction, Dorothy accidentally saved over the order in the Emerald system. To her horror, she realized that any other link in the system pointing to that order was now broken and giving an error message.

Panicked, Dorothy tried to recover the lost order. She spent over an hour searching through the system trying to undo her mistake, but to no avail. The order seemed to be gone, and with it, the details of the important shipment for Flying Monkeys Studio. Dorothy worked closely with the CEO over at Flying Monkeys Studio on many orders and knew her to be particularly wicked if things did not go smoothly.

Feeling a knot of dread in her stomach, Dorothy went to her supervisor, Glinda. She explained the situation, her voice shaky with worry. Glinda's face grew serious as she listened, and when Dorothy finished, she said, "Dorothy, that was a very important order. And you know how their CEO can be. You need to contact Emerald support immediately."

With a nod, Dorothy hurried back to her desk, picked up the phone, and dialed the number for Emerald support. As she waited for someone to pick up, she could only hope that they would be able to help her recover the lost order.

Oh no! What an unfortunate situation for Dorothy to be in. But the biggest issue here Dorothy is facing isn't a technical one. It's emotional. **This chapter is going to double down on some previous key lessons and reinforce a lot of what we've discussed so far.**

It wasn't until a couple of years into my career I learned that understanding my clients' pain—their needs, fears, and frustrations—was just as important as diagnosing and solving their technical problems, if not more so. By putting myself in their shoes and truly empathizing with their situation, I was better equipped to provide solutions that not only resolved the issue at hand but also addressed their underlying concerns.

In order to understand their pain, though, I had to understand and care about the companies they worked for and their role in those companies. These things went hand-in-hand. And how do you go about understanding and caring about those things? You do it by genuinely engaging with the people you support, actively listening, and taking an interest in them and their situation. If you want to level up your IT career, your relationship with users and clients can't just be transactional and scripted.

That's why I wanted to go a bit more in-depth with the premise of our story here. As we move forward, we'll see how Dorothy's interaction with Emerald's support team highlights the importance of empathy and rapport-building in IT support. We'll see how understanding a client's pain (their needs, fears, and frustrations) is just as crucial as diagnosing and solving their technical problems.

Pay Close Attention to the Man Behind the Phone

Dorothy, still shaken, dialed the support number for Emerald. The phone was answered by a man named Oscar. Oscar had a warm, reassuring voice that immediately put Dorothy at ease. He listened attentively as Dorothy explained her predicament.

"Oh dear, that does sound like a pickle," Oscar said sympathetically when Dorothy finished her explanation. "But don't worry, Dorothy. We'll get this sorted out."

Oscar's empathetic response and his assurance of support helped to alleviate Dorothy's stress. As Oscar looked through documentation in his system for Yellow Brick Transportation, he continued his conversation with Dorothy. "Dorothy, I'm not sure if we've ever spoken before. What is it you do over at Yellow Brick?"

Dorothy told Oscar about her job role. Oscar responded, "Oh, an order supervisor? Nice! I love working with you guys." Oscar then lowered his voice to a little above a whisper, speaking conspiratorially to Dorothy, "I'll let you in on a secret, this type of thing happens all the time. This is the first time I've ever had to talk to you and I see you guys have been with our system for about 2 years now, so trust me, you're ahead of the curve!"

Dorothy began to feel a bit more at ease, and Oscar had asked to get remoted into her computer. She helped him gain access, then asked, "Are you sure we can get this back?" Oscar lightened the mood again with a joke, saying, "Don't worry, Dorothy. We're not in Kansas anymore, but I'm sure we can find a way back home."

Oscar then guided Dorothy through the process of recovering the lost order. He was patient and clear in his instructions, making sure Dorothy understood each step. He also made sure to use simple language and analogies to explain the technical aspects of the recovery process.

"Think of it like a time machine, Dorothy," Oscar explained. "We're going to go back in time to before the order was overwritten."

After about 30 minutes, Oscar managed to recover the lost order entirely and they tested all the relational links together to verify they worked. Dorothy was relieved and thanked Oscar profusely. But Oscar wasn't done yet. He wanted to ensure that Dorothy would be prepared if such an issue arose in the future. So, he created a simple "How-To Guide" for recovering lost orders and sent it over to Dorothy.

"This guide is like a map," Oscar said. "If you ever find yourself in a similar situation, you can use it to find your way back."

Dorothy was grateful for Oscar's help and his proactive approach. She thanked him again and ended the call, feeling much more confident and less stressed. She realized that Oscar had not only fixed her immediate problem but had also equipped her with the knowledge to prevent similar issues in the future. To Oscar's credit, he took this sort of approach with all of the end users he helped on a daily basis.

A year later, Yellow Brick Logistics had grown and promoted Dorothy to Chief Logistics Officer. They began to acquire more firms and incorporating them into the fold required them to move from the Emerald software to another platform or to integrate Emerald into the new offices.

Because of her positive experience with Oscar, Dorothy contacted her account representative at Emerald and requested Oscar by name to be her dedicated account technician for the migration. Emerald software was happy to oblige, and they won a new 3-year contract with Yellow Brick Transportation. Because of the win, Oscar himself was promoted to Technical Account Manager where he oversaw the Yellow Brick Transportation account along with many others, and the partnership flourished.

See the power of empathy and rapport in action? Oscar, the support technician, demonstrates these qualities in his interaction with Dorothy. He listens attentively to her problem, reassures her, and guides her through the recovery process. He also lightens the mood with humor and (building off of previous lessons we've discussed) uses simple language and analogies to explain technical concepts. This approach not only helps to alleviate Dorothy's stress but also builds rapport and trust.

Oscar's proactive approach is also noteworthy. After resolving the immediate issue, he goes a step further to prevent similar issues in the future. He creates a "How-To Guide" for Dorothy, equipping her with the knowledge she needs to handle similar situations. This not only empowers Dorothy but also demonstrates Oscar's commitment to providing excellent customer service.

Dorothy's day is much better and her experience has left her feeling great about the relationship her company has with the software vendor. Oscar's approach really helped build up something called an "Emotional Bank Account."

The concept of the "Emotional Bank Account" is a powerful metaphor that encapsulates the essence of building and maintaining healthy relationships.

(It's a concept that was popularized by Stephen Covey in his book "The 7 Habits of Highly Effective People").

In the context of tech support, the Emotional Bank Account represents the trust and goodwill that you've built up with a client or user. Every positive interaction you have with them—every issue you resolve, every question you answer, every extra mile you go—makes a "deposit" into this account. Conversely, negative interactions, such as unresolved issues or poor communication, make "withdrawals". The more you deposit with clients and users, the more that positivity spreads to others, fostering more goodwill between you and the organization you're working for or the businesses you're partnered with.

In the story, Oscar made several significant deposits into the Emotional Bank Account with Dorothy. He listened to her, reassured her, solved her problem, and provided her with a tool to prevent future issues. These actions not only resolved the immediate issue but also increased Dorothy's trust and goodwill towards Oscar and Emerald.

However, it's important to note that maintaining a healthy Emotional Bank Account requires consistent effort. **One**

positive interaction won't make up for a history of negative ones. But consistent positive interactions can build a strong relationship that can weather the occasional negative experience.

Consistently providing this level of support is crucial to long-term growth.

In the end, the Emotional Bank Account is about building trust, demonstrating empathy, and providing effective solutions. By focusing on these areas, you can build strong, positive relationships with your clients or users, leading to better outcomes for everyone involved.

But here's an example of the real power of this lesson: the story highlights the long-term benefits of building an Emotional Bank Account through providing empathetic and effective support. Because of her positive experience with Oscar, Dorothy requested him by name when her company needed further support. This led to a new contract for Emerald and a promotion for Oscar. It's a win-win situation that underscores the value of empathy, rapport, and proactive problem-solving in IT support.

You might say, "But Aaron, this is just a story that you created! Of course it worked out great for the characters!" As I mentioned in the Preface, these fables are based on or drawn from my own real-world experiences. Since we've done a few of these stories now, I thought it would be a good time to reiterate this. This particular story highlights my own path to a Technical Account Manager role – not one specific instance with one specific client that led to a big win, but a consistent effort and mindset I adopted every day in previous roles to providing support for clients and users that consistently deposited into their Emotional Bank Accounts and, ultimately, the companies they worked for. Recognizing those efforts led to more wins, more business, and my name being mentioned positively by both peers and clients alike. And this, in turn, led to my own promotion and increased salary.

I WANT YOU TO MAKE MORE MONEY TOO

As we continue, remember that these fables are rooted in truth and the lessons within them represent real opportunities you can seize to grow and get to that next level.

💡 **KeyTakeaways from this Chapter**
- ✓ Technical fixes are important, but how you make clients and end users feel is even more crucial.
- ✓ Empathy, humor, and reassurance turn a stressful situation into a positive experience.
- ✓ Use analogies and simple explanations to build rapport and reduce frustration.
- ✓ Make deposits into the "Emotional Bank Account" by consistently delivering great service.
- ✓ Long-term career growth (and higher salaries) comes from being someone clients and end users trust and remember.

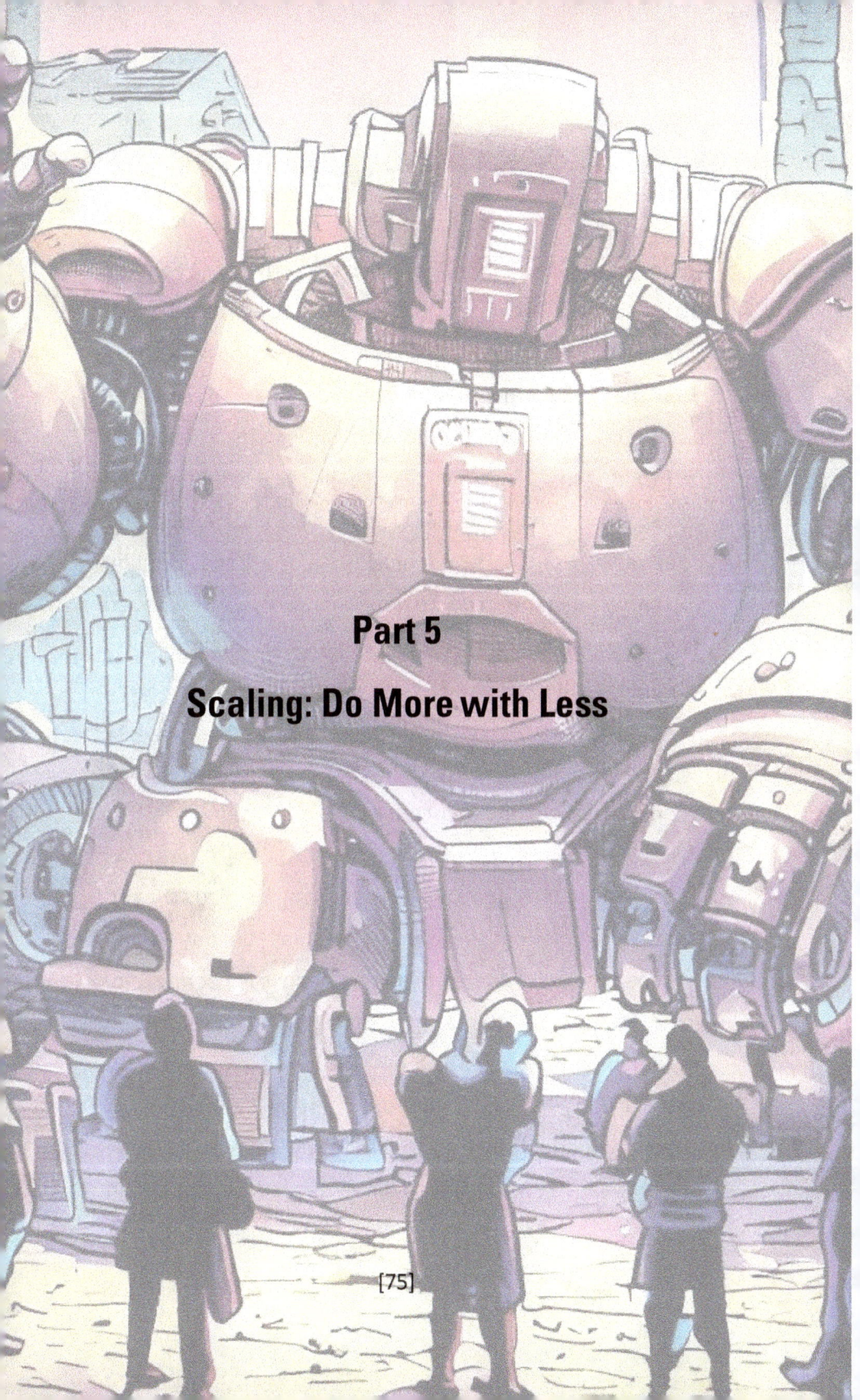

Part 5

Scaling: Do More with Less

Premise

Ben was a seasoned Systems Administrator at
Ultrabyte Services, a relatively small IT
managed services provider in the town of
Springfield. He had been working in the IT
industry for nearly 10 years now in a variety
of odd jobs, but mostly technical support and
helping to manage small business networks and
servers. Ben was naturally a curious person
with a desire to learn and grow. He loved
helping people, but the technical work was no
longer challenging. He yearned for more
efficient ways to deliver IT support. He was
tired of the repetitive tasks and the time-
consuming manual processes.

Ben was also growing increasingly frustrated
with the inefficiencies in his day-to-day work.
Ultrabyte was using a hodgepodge of technology
and manual processes to manage their customer
environments. This included manually
performing maintenance, applying Windows
updates and patches, manually checking for
performance related concerns such as high

memory usage or low disk space, and manually enabling features and installing software.

Ben had been thinking about it for a while now. He knew that the current way of doing things was not sustainable, especially as the company continued to grow. So, he decided to bring it up with Jacob, the Helpdesk Manager.

"Jacob, can I talk to you for a minute?" Ben asked one day, catching Jacob as he was about to leave for lunch.

"Sure, Ben. What's up?" Jacob replied, pausing in his tracks.

"I've been thinking about our current workflow," Ben began. "We're doing a lot of things manually, and it's taking up a lot of our time. I think we could benefit from some automation."

Jacob looked thoughtful. "I see where you're coming from. But implementing new software can be expensive. And there's always the risk that it might not work as expected."

"Absolutely. And I understand those concerns," Ben said. "But I believe the benefits outweigh the risks. With automation, we could save hundreds of man-hours. We could also reduce the chances of errors and improve our service quality."

Jacob nodded. "I'm open to the idea. But we'll need to justify the change. Can you put together some figures? Show me how much time we could save, how much it would cost, and how it would improve our services."

"Sure thing," Ben replied, feeling a surge of excitement. "I'll get on it right away."

The premise of this story highlights a common challenge in the IT field: work-related ennui and frustration due to repetitive tasks and inefficient processes. This is a situation that many IT professionals can relate to. It's not uncommon to feel unfulfilled or bored when the work no longer presents a challenge or when the processes in place seem outdated or inefficient.

Ben is a seasoned Systems Administrator who is feeling the pinch of these inefficiencies. He is yearning for more efficient ways to deliver IT support and is tired of the repetitive tasks and time-consuming manual processes. This is a sentiment that many IT professionals can resonate with. The desire to improve, to make things better, is a common trait in the IT industry.

What sets Ben apart, however, is his willingness to act on his feelings.

Instead of just complaining or feeling frustrated, Ben decides to do something about it. He brings up his concerns with his manager, Jacob, and suggests the idea of automation. This is an important lesson for anyone feeling unfulfilled or frustrated in their job: communication is key. It's important to be honest about how you're feeling and why, and to be willing to act on it.

Jacob, on the other hand, represents a common reaction to change: hesitance and concern about cost and results. However, he is open to the idea and asks Ben to justify the change with numbers. This is a realistic depiction of how many managers would react to such a suggestion. (It's important to note that being open to new ideas and willing to consider them is a crucial trait for any good manager.)

The premise sets the stage for a story about change, improvement, the power of taking initiative, and putting work in during your own time. It's a story that many IT professionals can relate to and learn from. It emphasizes the importance of communication, the willingness to improve, the courage to act on your feelings, and the willingness to sacrifice your own time to grow in your career.

The Midnight Oil

With a renewed passion, Ben set out to make a case for automation. He knew it wouldn't be easy, but he was determined to make a difference. He also knew he wouldn't be able to take time out of his workday to do the research. Customer demands were high, and the company was lean. Ben knew if he wanted to do this, he'd have to do it on his own time. But that didn't deter him.

In his spare time, Ben decided that he would start researching new technology and automation. He was particularly interested in Remote Monitoring and Management (RMM) platforms, which are designed to help IT service providers manage their client's IT systems more efficiently. He came across several platforms and created demo accounts for each. Over the course of several weeks, Ben dove into these technologies, read technical how-to documentation, configured and tested them on his own, and kept his thoughts in Microsoft OneNote.

After weeks of testing, a platform called "Insight" was the one that caught his attention the most. Insight was a comprehensive, cloud-based RMM platform that included a built-in ticketing system, which would allow Ultrabyte Services to streamline their support processes.

Ben saw the potential of Insight to transform the way Ultrabyte Services delivered IT support. He envisioned a future where repetitive tasks were automated, alerts were responded to proactively, and software installations were done with the click of a button. He knew that implementing Insight would not only save hundreds of man-hours but also improve the quality of support provided to their clients.

With this vision in mind, Ben set out to bring about this transformation at Ultrabyte Services. He knew it wouldn't be easy, but he was driven by the potential benefits and the opportunity to make a real difference in the company.

Ben calculated the hard costs of the software (a few dollars per device monitored per month). He also calculated the time saved performing certain tasks that he commonly did manually for the company. The benefit of testing the program for weeks allowed him to see just how long it took to do something with Insight vs. doing it manually. Knowing how much Ultrabyte charged per hour for non-contract customers, he was able to extrapolate soft cost savings by implementing Insight. Further, he was able to determine that Insight would allow them to forego hiring a new technician immediately and allow the device-to-tech ratio to scale from 100:1 to 200:1. By his math, although Insight would cost a few thousand dollars of new hard costs per month, Ultrabyte would save tens of thousands of dollars in the long run, and customers would have faster and more reliable service.

It was time to take his findings back to Jacob.

So, here we see Ben's determination and passion for improvement come to life.

Despite the challenges, he takes it upon himself to research and test new technologies in his own time. This is a testament to his dedication and his desire to make a difference. It's also a great example of taking the initiative and going the extra mile, traits that are highly valued in any professional setting. This is a necessary quality that separates highly successful technical people from mediocre ones.

Ben's focus on Remote Monitoring and Management (RMM) platforms is a nod to the importance of leveraging technology to improve efficiency and service delivery in the IT industry. RMM platforms, like the fictional one Ben chooses, "Insight", are designed to streamline IT support processes, automate repetitive tasks, and proactively respond to alerts. This not only saves time but also improves the quality of support provided to clients.

(Please feel free to research some actual RMM platforms if this is interesting to you; I highly recommend NinjaOne for what it's worth. You can find it at https://www.ninjaone.com).

The detailed research and testing that Ben undertakes highlights the importance of due diligence when considering new technologies or processes. It's not enough to simply have a good idea; it needs to be backed up by solid evidence and thorough testing. This is a crucial step in any change management process. It speaks to your thoroughness and your ability to see beyond just the bells and whistles of a cool piece of technology.

The most successful IT people are the ones who understand the intersection of business and technology.

Ben's calculations of the costs and potential savings associated with implementing Insight underscore the importance of making a strong business case when proposing changes. It's not enough to simply say that a new process or technology will improve things; you need to show how it will benefit the business in tangible terms. In this case, Ben is able to demonstrate that although Insight would incur some new costs, it would ultimately result in significant savings and improved service for Ultrabyte's clients.

I can't emphasize enough the importance of taking initiative, doing your due diligence, and making strong business cases when proposing changes or improvements. It's a lesson in perseverance, strategic thinking, and the power of technology to transform business processes.

Where the Rubber Meets the Road

With his research and calculations in hand, Ben approached Jacob for a follow-up meeting. He had prepared a detailed presentation, outlining the benefits of implementing Insight, the cost savings, and the improved service quality.

"Jacob, I've done the research and crunched the numbers," Ben began. "I believe implementing Insight will not only save us hundreds of man-hours but also improve the quality of our support."

Jacob listened attentively as Ben presented his findings. He was impressed with the depth of Ben's research and the potential benefits of implementing Insight.

"Your research is thorough, Ben," Jacob said. "I'm impressed. And the potential benefits are significant. I think we should move forward with this."

Ben was thrilled. "Thank you. I believe this will make a real difference for us and our clients."

Jacob nodded. "I agree. But implementing a new system is a big task. We'll need to ensure it's deployed properly and that everyone is trained on how to use it."

"I understand," Ben replied. "I'm ready to take on that responsibility. I've already spent a lot of time with Insight, and I'm confident I can help everyone get up to speed."

"Excellent," Jacob said. "Let's get started."

Over the next few weeks, Ben worked tirelessly after hours to implement Insight. He configured the system to automate tasks, set up alerts, and created scripts for software installations. He also developed a training program to help his colleagues learn how to use the new system. While Ben was tired and sacrificed a lot over the last several months, he had challenged himself, learned an incredible amount about various new technologies, was implementing something that would improve the company he worked for and their clients' experiences with the company, and felt more fulfilled than he had in a long time.

The transition was not without its challenges, but Ben's dedication and hard work paid off. He was proud of his accomplishment and was ready to go into work with renewed energy and passion for what he was doing every day. On top of that, he was now a go-to resource as a subject matter expert for what quickly became his company's main line of business software.

Now we see Ben's initiative and dedication come to fruition. His thorough research and detailed presentation to Jacob showcase his commitment to improving the company's operations. His willingness to take on the responsibility of implementing and training others on the new system demonstrates his leadership and dedication to the company's success.

This story is a testament to the power of taking initiative and seeking out opportunities for improvement in one's work. It's not uncommon for individuals in the IT field, or any field for that matter, to feel unfulfilled or bored with their work. However, as Ben's story illustrates, these feelings can be a catalyst for change and improvement.

Most technology professionals are perfectly content with resting on their laurels. While that might make your day job easy, it is not the path to growth. At best, you'll fly under the radar for years as an average employee making an average salary doing the same thing every day until retirement. At worst, you'll be let go from your place of work, with no relevant skills or experience to set you apart in the next job interview.

Ben didn't just accept the status quo, wasn't content with flying under the radar and collecting a paycheck. He saw an opportunity to improve the company's operations and took it upon himself to research and propose a solution. This proactive approach not only improved the company's operations but also allowed Ben to challenge himself and grow professionally.

Furthermore, the story highlights the importance of effective communication and persuasion in implementing change. Ben didn't just do the research; he also presented his findings in a compelling manner that convinced Jacob of the benefits of implementing Insight.

Finally, all of this underscores the importance of dedication and hard work in achieving one's goals. Implementing a new system and training others on its use is no small feat. It required a significant amount of time and effort on Ben's part. However, his dedication and hard work ultimately paid off, leading to significant improvements in the company's operations and his personal fulfillment.

Now, I know what some of you are thinking. Some of you are thinking, "What kind of pro-corporate, neo-fascist, Ayn Rand loving capitalist propaganda is this? Ben was happy to sacrifice months of free labor for the benefit of his boss laughing all the way to the bank with nothing but personal fulfillment to show for it?!" Read this in Billy Mays' voice: But wait, there's more! Our story isn't quite over yet 😉

What About the Cheddar?

A few weeks after the successful implementation of Insight, Jacob approached Ben. He was impressed with Ben's dedication and the seamless transition to the new system. He saw the potential in Ben and wanted to offer him a new role.

"Ben, I've been thinking," Jacob began, "You've done an incredible job with the implementation of Insight. I'd like to adjust your daily responsibilities to oversee the entire automation system."

Ben was thrilled. He had worked hard to bring about this change and was excited about the opportunity to continue improving the company's operations. However, he also knew that he had invested a lot of his personal time and effort into this project. He felt that his contribution should be recognized officially.

"Jacob, I appreciate the offer," Ben replied, "But I've spent a lot of my own time and effort making this happen for the company. I think it would be fair to create an official job role for this, complete with a salary re-negotiation."

Jacob nodded, "I understand where you're coming from. Let's come up with a job description together. How about Network Operations Manager?"

Ben agreed, "That sounds great. I'll do some research on competitive salary data and get back to you."

Over the next few days, Ben researched the average salary for a NOC Manager. He then presented his findings to Jacob. After some negotiation, they agreed on a new salary that reflected Ben's new responsibilities and the value he brought to the company.

Ben happily accepted his promotion and raise. He was excited about his new role and the opportunity to continue improving Ultrabyte Services' operations. His hard work and dedication had paid off, and he was ready to take on the challenges of his new position.

See? I told you the story wasn't over yet.

While job fulfillment is extremely important, and I 100% advocate for you to find fulfillment in what you do, the ultimate goal of this story wasn't just to show you how to overcome boredom or frustration in your job, it was to show how you can achieve serious job growth through the concept of deferred gratification.

(You can read about delayed gratification here: https://en.wikipedia.org/wiki/Delayed_gratification)

Ben could have taken the easy path. He could have worked his 9 to 5, put in as little effort as possible, earn his paycheck, and then go back home and post his exaggerated grievances on the Anti-Work subreddit about the dystopian hellscape of 21st century wage slaving from the comfort of his climate-controlled apartment. But he didn't. He worked hard, he sacrificed. He gained new skills, he convinced his company to implement software that he alone became a subject matter expert for at the company (making himself invaluable to his employer), and was ultimately given a raise and a promotion. It's not as uncommon as you might think, and I've done this very thing myself on more than one occasion.

Also, don't forget the title of this fable. By helping a company save money and scale, you can metaphorically carve yourself a slice of the savings pie and grow your own salary.

Obviously, implementing an entire RMM system is a monumental task, but I can illustrate the power of scalability (and some of our previous lessons too) with a much simpler example.

I recall a time where a user (let's call her Susan) of a relatively large client needed help getting connected to VPN. She called in often for this sort of thing and complained about how forgetful she is. One day, instead of just fixing the problem manually, I decided to create a How-To "Cheat Sheet" for using the VPN and I gave it to Susan. Even though it took me about an hour to make it (complete with screenshots), Susan was very grateful and she only ever had to call in for VPN issues if something unexpected happened. It was something that really deposited into the emotional bank account and ingratiated me with Susan and the company she worked for.

But there was an even larger, more beneficial way to go even further than I already had. I decided to use that document as a template. I worded it more generically than something super customized for Susan. I added our organization's logo to it in the header. I also created a small authorship credit to myself, so whoever I distributed the document to would know I was the one who wrote it. I then saved the document in a centralized location for the client and for our organization so other techs could benefit from what I wrote.

The customer service I provided was not just for one user. Or even just one organization. Because of one small client interaction, I spent 20 extra minutes making the company I worked for more scalable, and increased customer satisfaction at the same time. Every time a user reads that document in the future, it spares the company a 10-minute ticket. Every time a new tech needed to know how to set up VPN or remote desktop, I saved them 10 minutes of research. That 60 minutes of extra time I spent being thoughtful, actively listening, and wanting to provide great customer service wound up saving my organization hundreds if not thousands of dollars in soft costs over the span of years. And the document had my name on it to boot; everyone knew it was me that did it.

So, that is an extremely simple example of how even a level 1 desktop support tech can bring a ton of extra value to an organization and its customers. But, as you become more technically skilled, you can find other, more advanced ways to help scale your organization and reduce your own workload, just like Ben did. Scripting and automation are hugely beneficial to sysadmins and managed services providers. The more intimately familiar you become with something like PowerShell or bash scripting, the more valuable you will be for an organization.

Here's another example: Imagine working in-house for a hospital that has a tedious manual process for user onboardings and offboardings. You have to remote into a domain controller, find a user to copy, copy them, set a password manually, and manually check that they are part of a specific set of security groups in order to gain access to mapped drives on a file server. Then, you have to log into a laptop manually as an admin, and manually install all the software they need for the computer: productivity software such as Office 365 apps, probably an EHR (electronic health records) app, maybe VPN, etc. Then you have to log off the admin and login as the user, map drives manually (or maybe you're lucky enough that they are mapped through group policy) as well as add network printers and test everything. By the time you're done with this one new user, 2 hours or more has gone by from your day and you look to see 12 new helpdesk tickets that have gone unanswered. Looks like it will be another late night.

But what if you took a different approach? Don't spend 2 hours doing all that manually. Spend 4-6 hours figuring out how to script most if not all of that. Use Google or an AI assistant like ChatGPT and research PowerShell commands (**be careful with data exfiltration and always follow security policies**). Do test runs, trial and error. Painstakingly make a script that checks whether or not a user exists with that username, if not create it, automatically add it to XYZ distribution groups based on a department variable you enter. Make separate scripts to automatically download and install software, set up a VPN connection, map network drives and printers.

Do it on your own time if you have to, just as Ben did. After 2-3 user setups, all your time invested will be made up. Offer to show your technical colleagues how it functions after you successfully test it. Let them use it as well. Heck, build in a shoutout to yourself that flashes in the PowerShell window when the script executes.

By thinking like this and putting it into action, you can fast-track yourself for promotions and raises.

"But what if they fire me and benefit from my free labor after I put in all that extra time, effort, and energy?"

Remember, you're not just scaling the company you work for when you do this, you're scaling yourself. Any boss or place worth working for isn't going to want to lose an asset who goes out of their way to save them money through scaling. And, in the unlikely event that something like that did happen, it's a blessing in disguise. But to explore this point further, let's take a look at an alternate ending to this fable...

Alternate Ending: A Slap in the Face

A few weeks after the successful implementation of Insight, Jacob approached Ben. He was impressed with Ben's dedication and the seamless transition to the new system. He saw the potential in Ben and wanted to offer him a new role.

"Ben, I've been thinking," Jacob began, "You've done an incredible job with the implementation of Insight. I'd like to add to your daily responsibilities to oversee the entire automation system."

Ben was thrilled. He had worked hard to bring about this change and was excited about the opportunity to continue improving the company's operations. However, he also knew that he had invested a lot of his personal time and effort into this project. He felt that his contribution should be recognized officially.

"I appreciate the offer," Ben replied, "But I've spent a lot of my own time and effort making this happen for the company. It's too much to do on top of my day job. I think it would be fair to create an official job role for this and transition me full-time to that, complete with a salary re-negotiation."

Jacob paused, looking a bit uncomfortable. "Ben, I understand where you're coming from," he said slowly, "But the budget is tight right now. We unfortunately can't afford to give you a raise or compensate you for the work you've done for this. You did this of your own volition, we didn't ask you to do it."

Ben was taken aback. He had put so much effort and time into this project, expecting that his hard work would be rewarded.

Jacob continued, "But I promise you, Ben, if you continue putting in the effort, we'll look into giving you at least a small bonus at next year's performance review."

"Jacob, I understand the situation," Ben replied, masking his disappointment. "No problem."

However, as he walked away from the conversation, Ben couldn't help but feel a strong sense of dissatisfaction. While he knew there was a chance something like this could happen, he hoped that his hard work would be rewarded. But the company he had dedicated so much of his time to was unwilling to recognize and reward him for the value he brought.

Ben decided he was not content working for a company that didn't value his contributions. He spent the next few days updating his resume, showcasing all the new skills he had learned, his expertise with Insight, and how he had saved Ultrabyte a significant amount of money.

He applied for a Network Operations Manager position at several rival firms. One of them, a well-known competitor in the market, reached out for an interview. During the interview, Ben learned that they used Insight themselves and were thrilled to find someone with his level of expertise and passion. They offered him a substantial salary increase and a position managing their Insight platform.

With a new opportunity on the horizon, Ben went back to Jacob to give his two weeks' notice. Jacob was shocked and reacted with frustration.

"Ben, this is selfish," Jacob accused, his voice rising. "After all we've done for you? You're the only one who knows how to manage this new system and, if you leave, you'll be hurting the company!"

"Jacob," Ben replied calmly, "I have to do what's best for me. If Ultrabyte doesn't recognize my value, I have to take my loyalty elsewhere."

With that, Ben walked away, ready to embark on a new journey where his skills and contributions would be recognized and rewarded.

In this alternate ending, we see a different outcome for Ben. Despite his hard work and dedication, his efforts are not rewarded by his employer. This is a scenario that many employees may find themselves in. They put in extra effort, go above and beyond their job description, and yet their contributions are not recognized or rewarded.

However, Ben's story doesn't end there. Instead of accepting the situation he found himself in, Ben takes control of his career. He updates his resume, applies for new positions, and lands a job at a company that values his skills and expertise. This is a testament to the power of self-advocacy and taking control of one's career.

The two contrasting endings also highlight the importance of recognizing and rewarding employees for their contributions. When employees feel valued, they are more likely to be engaged and productive. On the other hand, when employees feel undervalued, they may start looking for opportunities elsewhere.

In the end, Ben's story is a lesson in resilience and self-advocacy. Despite the challenges he faced, he remained committed to his goal of improving his company's operations. When his efforts were not recognized, he didn't give up. Instead, he took control of his career and found a company that valued his contributions. In this alternate ending, Ben's efforts may have benefited Ultrabyte too, but that is just a by-product of Ben benefiting himself from the work he did. Ben didn't "donate his labor" to Ultrabyte, he donated his labor to himself. One way or the other, Ben was willing to defer his own gratification to be rewarded in the end. And, with Ben's departure, Ultrabyte is suffering, with no one skilled enough to continue managing the system after Ben leaves.

Whether you find yourself in a situation similar to Ben's original ending or the alternate one, remember that your value does not decrease based on someone's inability to see your worth.

Always be ready to advocate for yourself and your contributions. And remember, scaling isn't just about improving the company you work for; it's about improving yourself as well.

💡 **Key Takeaways from this Chapter**
- ✓ **Look for inefficiencies** in your work and find ways to streamline and automate.
- ✓ **Take initiative.** Don't wait for someone to assign you a project; proactively improve processes.
- ✓ **Frame solutions in business terms.** Show how they save time, reduce costs, and increase efficiency.

- ✓ **Invest in yourself.** Learning new skills outside of work can lead to career advancement.
- ✓ **Deferred gratification pays off.** Short-term effort can lead to long-term promotions and salary increases.
- ✓ **Make yourself indispensable** by becoming the **go-to expert** for a critical system or process.
- ✓ If your value isn't recognized, **advocate for yourself.** Negotiate your worth or find a place that will.
- ✓ **Scaling isn't just about the company.** It's about making **your own work** more **impactful** and **future-proofing** your career.

Part 6

Are You Strengthening or Weakening

Your Team?

Premise

Before we dive into another fable, I want to ask myself the question at the heart of this section: Am I strengthening or weakening my team?

You and I are on a kind of team: you as the reader, and I as the writer. You're reading this book for guidance or mentorship, and I've written it to help you grow. But what if my approach to storytelling hasn't been effective? What if this lesson would be better learned with a different approach?

As the leader of this team, it's my responsibility to reflect on how my leadership and communication impact you—the other team members.

So, let's examine what it means to be part of a team and how you can contribute in a meaningful way (or, at the very least, avoid becoming a liability).

The most successful technology professionals and organizations are those that harness the power of resourcefulness, collaboration, teamwork, and effective communication.

As a trusted technical or security resource, your role extends beyond being a technical expert. You must be a unifying force that brings people together, motivates them, and helps them grow (even if you are not in a position of management). The strength of your team can make or break your (or your organization's) ability to adapt, innovate, and stay competitive in whatever industry your organization is a part of.

It's crucial to recognize that each team member can bring unique skills, experiences, and perspectives to the table. Embracing this diversity of thought and fostering a culture of inclusion can also lead to more innovative solutions and ultimately drive your team's success. Moreover, in the fast-paced world of technology, it's essential to cultivate a learning mindset for yourself and among your peers, encouraging them to continuously expand their skillsets and stay up-to-date with industry trends.

There's a delicate balance in teamwork...

- ➤ Seek support when needed. Ask for help and receive coaching from senior staff.
- ➤ But don't depend on others to do your thinking for you.

Before asking for help, show resourcefulness…

- ➤ Try solving the problem first.
- ➤ Do your research.
- ➤ Demonstrate initiative.
- ➤ Ensure it's clear you've already put in effort before seeking guidance.

This is all especially true for relatively simple problems or those that can be solved with a little elbow grease and critical thinking (regardless of whether the problem is technical or non-technical).

Your actions and decisions can either strengthen or weaken your team. While high standards and accountability push people to excel, admonishment or shaming can stifle creativity and hinder collaboration. A balance must be struck between providing input and support, and allowing team members the autonomy to take risks, learn from their mistakes, and grow professionally.

Ask yourself: **"Are my actions and decisions strengthening or weakening my team?"** By reflecting on your own contributions to your team or peers and adopting the principles and strategies I provide, you can play a vital role in transforming your team or organization into a powerhouse of innovation, adaptability, and excellence, paving the way for enduring success and unlocking the next level of your career.

Now, let's see what happens when arrogance, poor self-awareness, and an unwillingness to adapt don't just weaken a team—they push it toward collapse.

Sam was a software developer at Scalesoft, a mid-sized software development company that specialized in creating custom solutions for a variety of clients. They were part of a team of developers, each with their own strengths and areas of expertise. Sam's experience was mostly from supporting and coding for smaller organizations, which gave them a certain level of confidence in their abilities. They loved to talk about their coding skills, often boasting about how great they were and how much all of Scalesoft's customers loved them. Sam's overconfidence was more than just a quirk. It was part of their reputation.

"Scalesoft had just landed a major client: Yamazaki Industries, a global manufacturing giant based in Tokyo. The sales team had fought hard to win the contract, and now the entire company was buzzing with excitement. The project was to develop a custom integration between Yamazaki's Customer Relationship Management (CRM) software and a handful of their other existing line of business systems. The integration was also expected to provide advanced features with the CRM itself such as predictive analytics, customer segmentation, and automated marketing campaigns.

The project was a significant one, with specific requirements and a tight deadline. It was a challenging task that required the best efforts of the entire team. When discussing who would be part of the project team, Sam had volunteered.

"I got this. Don't worry," Sam insisted, too confidently.

In our new fable, we meet Sam. Sam is a character that many of us might recognize from our own workplaces (or, perhaps, you are Sam). They are confident, perhaps overly so, and have a reputation for boasting about their abilities. This inflated sense of self-worth, while it might seem harmless or even amusing at first, can often lead to issues down the line, especially in a team setting.

The company is embarking on a high-stakes project with a tight deadline, demanding the best from the entire team.

When the time comes to assemble the project team, Sam is quick to volunteer, insisting that they've "got this." This is a defining moment. Stepping up is admirable, but is Sam stepping up for the right reasons, or is their overconfidence blinding them to the risks? Are they truly prepared for the challenges ahead, or are they setting themselves, and their team, up for failure?

As we delve into this, we'll explore the consequences of Sam's actions and attitudes, and how they impact not only the project but also the dynamics within the team. We'll see how unchecked arrogance, disregard for teamwork, and a failure to be resourceful can lead to an inevitable downfall.

A Self-Aggrandizing Kickoff Meeting

The project kickoff meeting was held in the main conference room at Scalesoft. In attendance were Amit, the project manager; Ravi, the database administrator; Supriya, the senior developer; and, of course, Sam, a junior developer. The air was thick with anticipation; everyone understood how high the stakes were.

Amit stood at the front of the room, setting the stage. "Yamazaki Industries wants us to integrate their CRM software with their existing systems," he explained. "They also need advanced features like predictive analytics, customer segmentation, and automated marketing campaigns. This is a complex project with a tight deadline, so we need to stay focused and work efficiently as a team."

Before he could continue, Sam leaned forward and cut him off. "Oh, I've done something like this before," they said, flashing a smug grin. "I've already got some code we can use."

Ravi, trying to keep the discussion productive, responded evenly. "Sam, confidence is good, but this is a team effort. We all have different responsibilities, and this might be more complicated than you think."

Sam waved him off and launched into an unsolicited story about a past project, brimming with irrelevant details and incorrect technical jargon.

"This reminds me of when I helped a local retail shop set up their e-commerce website," they said, leaning back in their chair. "It was a WordPress site, and they needed it to sync with their inventory system."

The room fell silent.

Sam, oblivious to the tension, kept going. "I built a custom plugin for them. Piece of cake. I mean, it's basically the same thing we're doing here, right?"

Supriya and Ravi exchanged a glance. The comparison between a simple WordPress plugin and a complex enterprise CRM integration was so wildly off-base that neither of them knew how to respond at first.

Finally, Supriya cleared her throat. "Sam, while it's great that you've worked on integrations before, this is an entirely different level of complexity. We're dealing with multiple systems, real-time data processing, and advanced automation. It's not the same thing."

Sam dismissed her with a lazy wave. "Yeah, yeah, I get it. But coding is coding, right? How different could it be?"

Supriya clenched her jaw but let it go. Ravi exhaled sharply, giving Amit a look that said Do something.

Amit, sensing the tension, refocused the meeting. "Sam, we appreciate your enthusiasm, but let's stay on track. As I already explained, the client's requirements include predictive analytics and customer segmentation. These aren't simple add-ons, they require careful planning and execution."

Sam barely looked up from their phone.

After a long, uncomfortable silence, they finally glanced up. "So… what kind of features does the client want in the CRM?"

Amit's patience snapped. "Sam, I just explained that. Twice." His tone was even, but the irritation was unmistakable. "This is a critical project, and we need everyone to be fully engaged."

Ravi cut in before things could spiral. "I think it makes sense to build a formal project plan and start assigning tasks."

Supriya nodded. Amit, after another slow breath, did the same.

Sam just blinked.

The meeting dragged on, but by the time it ended, the unease in the room was undeniable. Despite their misgivings, the team had no choice but to give Sam responsibilities. And as they filed out of the conference room, a single thought lingered between them:

This project is too important to fail… but we just put part of it in the hands of someone who doesn't seem to grasp what's at stake.

Well, that was a disaster! The kickoff meeting for the project was a clear demonstration of the challenges that can arise when a team member lacks the necessary soft skills for effective collaboration. Sam displayed several behaviors that were detrimental to the team's dynamics and the project's success.

Firstly, Sam's interruption of Amit, the project manager, during his outline of the project's goals was a clear sign of disrespect and a lack of professional etiquette. Their boastful claim of having done something similar before, despite the project's complexity, showed an inflated sense of their own abilities and a lack of understanding of the project's scope.

Secondly, Sam's dismissive attitude towards their team members' input, particularly Ravi's attempt to bring the conversation back on track, was concerning. Their insistence on comparing a simple WordPress plugin to a complex CRM integration project demonstrated a lack of technical understanding and an unwillingness to listen to their team members' expertise.

The most successful technical people know what they don't know and have the confidence to admit when they are out of their depth and need help or need to do research.

DON QUIXOTE CHARGING AT WINDMILLS

"TRUST ME, I KNOW WHAT I'M DOING!"

This is an instance where "fake it 'till you make it" is really bad and will set you and your team up for failure. Sam's attitude is a perfect example of the 'Don Quixote Effect,' blindly charging ahead, assuming confidence alone will lead to success. But in reality, this approach sets both the individual and their team up for failure.

Thirdly, Sam's lack of attention to (or incompetence about) the details, as evidenced by their question about the project's requirements that Amit had already answered, was a red flag. This lack of focus and disregard or ignorance for the importance of understanding the project's requirements could lead to significant issues down the line.

Finally, Sam's lack of professional behavior, such as getting distracted by their phone during the meeting, further demonstrated their lack of respect for their team members and the project as well as improper prioritization. It doesn't matter if it was a personal text message, a work email from another colleague, or a pop-up notification about a company announcement. Sam's attention should be on the project and their peers in the meeting.

Always give your full attention and skill capacity on meetings to your peers and clients.

By the time the meeting ended, the team's concerns had only grown. Sam's lack of collaboration, disregard for their teammates, and failure to grasp the project's complexity weren't just frustrating, they were a liability. And as we'll see, those concerns were well-founded.

Completely Out of Depth

The project was now in full swing, and the team was hard at work. However, Sam was proving to be more of a hindrance than a help. Despite their earlier boasts, they were struggling with the complexity of the project. Their work on the predictive analytics feature was riddled with bugs, and their notes were often unhelpful and lacking in detail.

One day, another peer, Jason, found Sam hunched over their computer, squinting at the screen. "What are you working on, Sam?" he asked.

Sam looked up, a hint of frustration in their eyes. "Just trying to figure out this predictive analytics thing," they said. "I'm reviewing some old WordPress plugin code I wrote a while back. I think it might help."

Jason raised an eyebrow. "Sam, is this for the project you're assigned? I think it's a lot more complex than a WordPress plugin. You can't just apply the same principles and expect it to work."

Sam shrugged. "Coding is coding, right? I'm sure I'll figure it out."

Later that day, Sam ran into another colleague in the break room, Ngai. "Hey Ngai, can you do something super quick for me? I know you're good with CRMs and I'm running around for all this work I have, it's just a coding thing." Sam asked.

Ngai responded, "Ummmm, OK. What's up?"

Sam grinned and replied, "Just need a predictive analytics plugin for this app. No big deal."

Ngai furrowed his eyebrows before responding, "Isn't that what you're responsible for doing on the project? Why are you working on other things and not focusing on that?" Sam replied, "No, no, no, this is for something else. I just thought you could give me some pointers."

Ngai sighed as Sam pulled him into a 45-minute conversation filled with basic coding questions. They were things Sam should have already known. Ngai tried to help but grew increasingly frustrated. It was clear Sam wanted someone else to do the work for them. Ngai eventually interrupted Sam and told them that he needed to grab the last 15 minutes of his lunch break to eat before getting back to his own work.

Later that day, Amit saw Sam hunched over their phone in the hallway. "Hey, Sam, what are you up to?" Sam looked up, startled. "Oh, just doing some research," they said, quickly locking their phone. Amit caught a glimpse of the screen before it went dark. It was a Google search for "how to do CRM coding."

Amit sighed. "Sam, you can't just Google your way through this project. You need to understand the specific CRM platform we're using and the coding language it requires. Do you even know what the name of the CRM application is or the coding language it's meant to be in? We've gone over this on multiple meetings."

Sam lied to Amit. "Oh, no, no, this is for something else. I haven't had a chance to focus on the predictive analytics module just yet, I had to wrap up all these other things I was roped into. You know how Zek Enterprises can be, but I'm their go-to. I've been working like crazy, you know I'm crazy good like that." Amit had no words and simply nodded before walking away, "OK, Sam. We'll talk later."

Over the next several days, Sam continued this trend of deception, lack of resourcefulness and respect, and trying to get other colleagues to do their thinking and work for them.

Amit was growing increasingly concerned. Sam's inability to grasp the complexity of the project, their ineffective methods for figuring things out, and their deceptive requests for getting others to do their work for them were all alarming as the deadlines began to loom.

Several more days passed by. Sam consistently avoided responsibility, either ghosting the team or offloading their work onto others, only to backpedal when caught. This type of teamwork was not only frustrating for everyone else but was also impeding the project severely. What little code Sam did complete turned out to be extremely buggy or incomplete, and their vague project notes on the documentation platform left the team in the dark about meaningful progress. This lack of transparency was causing the timeline to become strained, and morale was plummeting.

Amit decided to confront Sam about their performance. "Sam, we need to talk," Amit said, his tone firm. "Your work isn't just below expectations. It's actively hurting the team. Your code is riddled with bugs, your documentation is useless, and you've spent more time avoiding work than actually doing it. If you don't step up immediately, there will be consequences."

Sam shrugged, dismissing Amit's concerns. "I don't get why you're all ganging up on me," Sam scoffed. "I'm doing my best, but you just don't understand how I work. If you guys are gonna keep micromanaging me, maybe I should just quit and let you handle it yourselves." Amit was taken aback.

Despite Amit's confrontation, Sam's performance did not improve. Their code continued to be buggy, updates remained vague, and their attitude remained combative or dismissive. The project began to falter, and the team's concerns about Sam were becoming a reality. Their inflated sense of their own abilities and poor collaboration were leading to the downfall of the entire team and the project. Amit, Supriya, and Ravi were left wondering how they were going to shoulder the burden that Sam was forcing them to carry.

Alright, let's dissect what's going on here. Sam, our supposed coding prodigy, is proving to be more of a coding catastrophe. Sam's work was more bug-infested than a cheap motel mattress, and their notes made about as much sense as a scrambled TV signal at 3 AM. But let's not just point fingers at Sam, let's learn from them.

First off, look at Sam's approach to the project. They're trying to tackle a complex predictive analytics feature by drawing parallels with an old WordPress plugin they'd written. Now, I'm all for using past experiences to guide current work, but this is like trying to fix the Starship Enterprise with a bicycle repair kit. The lesson? Know the complexity of your task. Don't assume all problems are created equal. And most importantly: be radically honest with yourself and your team about your limitations. Whether it's coding, cybersecurity, cloud administration, or rocket science, self-awareness is key.

Next, we see Sam trying to offload their work onto Ngai under the guise of asking for help. Now, collaboration is key in any team, but there's a difference between asking for help and asking someone to do your job for you. Remember, your responsibilities are your own. Don't shirk them, embrace them. Even if that means you have to put in more time and effort to learn.

Then we have Sam's 'research' method – an uncomprehensive Google search. Now, Google is a fantastic tool, but it's not a substitute for genuine understanding. If you're tasked with working on something very specific and very complex, take the time to understand it. Don't just do a simple Google search and make it up as you go along. Read and process as much as you can to level up and become the expert you need to.

Your ability to Google-fu and be resourceful will only ever be as good as your ability to absorb and understand the work you're trying to do. This is what make's Sam's attitude and self-inflated ego even more annoying. Not only does Sam lack the skills they gloat about, Sam also lacks the potential or capacity to have the skills they gloat about.

(Whatever you do, don't become the poster child for the Dunning-Kruger effect...
https://en.wikipedia.org/wiki/Dunning-Kruger_effect)

And listen, I get it... just because Sam lacks the skills or the potential to be an exceptional coder doesn't mean they deserve to be treated badly or to not grow in their career. But it could mean Sam is not following the right career path. Knowing what you're good at and what you're not good at will go a long way to finding the right path of growth for your career.

However, perhaps the most concerning aspect of Sam's behavior is their lack of transparency and accountability. They're not keeping the team updated with clear, concise notes, and they're not taking responsibility for their work. This is a surefire way to lose the trust of your team and put any project or relationship at risk. It's not any one thing Sam is doing here, it's all of these things in aggregate and in context to each other.

So, what can we learn from this tale of woe so far? Understand the task at hand, take responsibility for your work, collaborate effectively, and strive for genuine understanding, not just quick fixes. And most importantly, be transparent and accountable in your work. Remember, we're not just pointing out Sam's mistakes for the sake of it. We're learning from them. So, take these lessons to heart and you'll avoid becoming the Sam of your team.

Be the kind of teammate others can count on. And most importantly, take feedback seriously. You don't have to agree with every criticism, but you owe it to yourself to at least understand where it's coming from. The difference between staying mediocre and becoming great is how well you learn from your mistakes.

It Comes to a Head

The project was in disarray. Sam's contribution (or lack thereof) had left a gaping hole in the team's work. The predictive analytics feature, once a promising aspect of the project, was now riddled with bugs and half-finished code. The team wasn't just behind schedule—they were drowning.

The once-lively office had become a war room of frustration. The team pulled late nights, fueled by caffeine and desperation, trying to salvage what they could. The deadline had already come and gone, but the project was still unfinished.

Supriya stared at her screen, fingers flying across the keyboard, eyes scanning line after line of Sam's broken code. "This is a disaster," she muttered under her breath. "It's like they've never written a line of code in their life."

Across from her, Amit let out a weary sigh, rubbing his temples. "I know," he replied, voice heavy with exhaustion. "I've never seen anything like this. It's like they didn't even try."

From his desk, Ravi shook his head. "We're all pulling double shifts because Sam couldn't be bothered to even put in an effort."

"And the worst part?" Supriya added, still typing. "We're the ones who will take the fall for this. The client doesn't care about our internal issues. They just want their project delivered."

The room fell silent. They all knew it was true.

It took another full week of exhausting overtime before the team finally finished the project. But the damage was done. As Supriya had predicted, the client was furious.

Amit was called into a meeting with the higher-ups. When he returned an hour later, his face was grim.

He wasted no time. "Yamazaki Industries is refusing to pay," he said flatly. "They lost a major contract bid because of the delays. They're holding us responsible."

The weight of those words settled over the room. No one spoke.

"And," Amit continued, "they're cutting ties with us completely."

Ravi exhaled sharply. Supriya closed her eyes. The exhaustion of the past few weeks now carried an even heavier burden. They had worked themselves to the bone, but none of it mattered.

Nothing more needed to be said. Everyone went home that night, drained and defeated, uncertain about what came next.

The following day, Sam was called into Leah's office. For once, the smugness was gone, replaced by quiet apprehension.

Thirty minutes later, they emerged, pale and wide-eyed. Without a word, they packed their things and left.

Word spread fast. Sam had been fired.

Shortly after, Leah gathered the rest of the department. She kept her tone professional but firm.

"As you all know, we had to let Sam go today. Their lack of accountability and unwillingness to collaborate cost us a client and hurt our reputation. I know the last few weeks have been incredibly difficult, and I want to thank you all for your hard work.

"But we have a long road ahead of us. The next few months will be challenging, and we'll need to rebuild, both as a team and as a company. I trust that we will."

Her words were measured, but the underlying message was clear: this situation would not be allowed to happen again.

Despite the long road ahead, there was an unspoken sense of relief in the room. Sam's incompetence had cost them time, money, and morale. Their removal, while overdue, was a necessary step forward.

For the first time in weeks, the team could breathe.

Sam had plenty of time to reflect. Programming had always seemed like the right path, but deep down, they had struggled with even the basics. Maybe this wasn't the right career after all.

Eventually, they took a job as an apprentice with a local plumber. To their surprise, the work felt fulfilling, hands-on, structured, and rewarding in a way coding never had been.

For the first time in a long time, things started to click. In the end, it was for the best. For everyone.

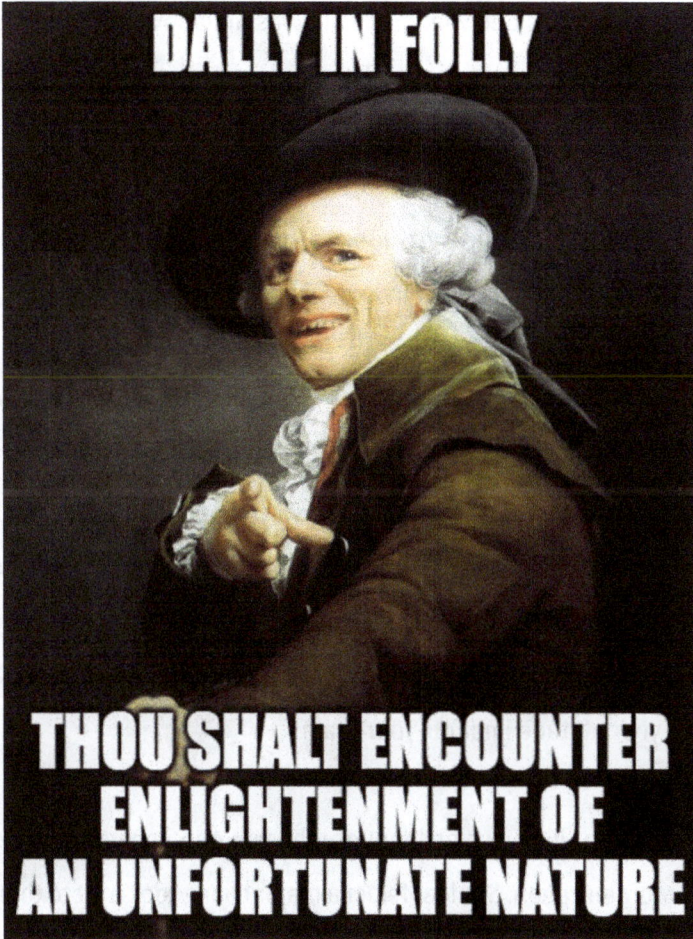

DALLY IN FOLLY

THOU SHALT ENCOUNTER ENLIGHTENMENT OF AN UNFORTUNATE NATURE

Sam's actions finally came back to bite them. And the fallout was ugly. Just like in real life, incompetence always has consequences, whether we like it or not. And it isn't always pretty. For Sam and Scalesoft, it's bad news for everyone involved, all around.

The project was delayed, the client was unhappy, and the team was left to pick up the pieces. This is a triple whammy! Ouch. The story also points out that clients won't care about your internal issues, they just want their projects delivered on time and up to standard. And when that doesn't happen, you can bet your bottom dollar that there will be consequences.

Just take a look at team morale, for example. Now, I don't know about you, but I've had to work overtime to fix someone else's mistakes in the past, and I was pretty miffed (putting it mildly). Frustration. Exhaustion. Resentment. The team was used to delivering high-quality work, but now they were stuck cleaning up Sam's mess, through no fault of their own. Your performance and your attitude impacts your team.

And it's not just your colleagues who you can impact, it's also your organization. Scalesoft lost a major client, took a hit to their reputation, and had to deal with a demoralized team. That's not just a bad day at the office, that's a full-blown crisis. And all because of one person's incompetence. Your actions and behaviors can have a ripple effect. In this case, the client refused to pay and even severed ties with the company. Sam was fired and left the company with their coworkers practically hating them. Scalesoft's reputation (and Sam's for that matter) are tarnished. And reputation is vitally important to success. It's like a credit score for your professional life. Once it takes a hit, it's hard to recover.

Think about it. Will Yamazaki Industries ever recommend Scalesoft to anyone? Will Amit, Supriya, Ravi, Ngai, Jason, Leah, or anyone else at Scalesoft give a recommendation for Sam? Not only will Scalesoft not be recommended, but Yamazaki may actively push partners to avoid Scalesoft. As for Sam, they'll be hard-pressed to list anyone at Scalesoft as a good reference.

The takeaway: your actions influence everyone around you, and those effects can extend far beyond your own little bubble. It's up to you whether or not you have a net positive or net negative effect. Be competent, be accountable, and be aware of the impact of your actions. And remember, if you're not cut out for the job, it's okay to admit it. It's better to find a career that suits you than to stick with something you're not good at. Because in the end, the only person you'll end up fooling is yourself.

As you navigate through your career, strive to be the best you can be, to work well with your teams, and to be aware of the impact of your actions. And hey, if tech isn't for you, that's okay.

Not everyone is cut out for this world. Find where your strengths truly lie.

However, a tale of failure alone, no matter how poignant the purpose, doesn't fully capture the lesson. Obviously (I hope) it's apparent that you avoid failing your team. But what's the contrapositive to that lesson? It's how to contribute to your team's success. Ask yourself, "**How can I cultivate a positive, successful team experience?**"

Foster open communication: Encourage the sharing of ideas, feedback, and knowledge among team members.

Creating an environment where team members feel comfortable sharing their ideas, feedback, and knowledge is essential for effective teamwork. Encourage open and honest dialogue, making sure everyone has an opportunity to contribute. By doing so, you'll foster a sense of ownership and shared responsibility among your team members, leading to better results and stronger relationships. If you're a novice team member, don't be afraid to raise your hand or offer insight. Also remember to ask questions!

I've seen how open communication can lead to breakthroughs in challenging situations. In one project, my team was working on a particularly complex system integration. There were various opinions on how to tackle the project, but initially, not everyone felt comfortable voicing their thoughts. I initiated a brainstorming session where each team member could openly share their perspective. One junior team member, who initially hesitated to speak up, suggested an innovative approach that ended up being the key to our success. By fostering open communication, we were able to unlock the full potential of our team and achieve our project goals.

Be an active listener: Pay attention to your colleagues' input and acknowledge their contributions.

Active listening goes beyond merely hearing what your colleagues are saying, or even the intellectualization of empathy—it involves genuinely understanding their point of view and acknowledging their contributions. By actively engaging with your teammates and asking clarifying questions, you demonstrate that their input is valued and encourage open communication.

I recall a time when a colleague of mine, Jess, was presenting her solution to a particular challenge during a team meeting. At first, her idea seemed counterintuitive, and many team members were quick to dismiss it. However, I decided to practice active listening and asked Jess to explain her thought process in more detail. As she did, it became clear that her idea had merit and could save us time and resources. By actively listening and giving Jess the chance to fully express her thoughts, our team was able to embrace her solution and achieve a better outcome.

Offer support and assistance: Help your teammates when they need it, and be willing to ask for help when you need it as well.

Providing help when your teammates need it and being open to receiving support when you need it are critical components of effective teamwork. By offering assistance, you build trust and collaboration, creating a positive environment where team members feel empowered to contribute their best work.

A few years ago, I was working on a project plan that involved a tight deadline and multiple dependencies. One of my teammates, Chris, was struggling to keep up with his tasks.

[142]

Instead of letting him flounder, I offered to help him with some of his workload. In return, Chris helped me with a different aspect of the project plan, leveraging his unique skill set. Our mutual support allowed us to complete the project plan on time, exceed client expectations, and helped the sales team close a deal. This experience reinforced the importance of offering and accepting assistance in a team setting.

Embrace diversity: Recognize the value of having team members with different backgrounds, perspectives, and skill sets.

Diversity in a team is a valuable asset. When team members come from different backgrounds, possess diverse skill sets, and offer varying perspectives, they can provide unique insights that lead to innovative solutions. As a team manager, it's important to keep that in mind. As a young tech professional, you can also benefit from being on a team with multiple perspectives.

Earlier I mentioned working with a team member named Jess, who brought valuable insights to our projects. One memorable instance involved a non-profit partner of ours that focused on victim advocacy and provided protection against domestic violence and sexual assault.

Our team was responsible for developing cybersecurity and information security policies for the non-profit. We initially provided them with our standard company-written policies, which were generally sufficient for most organizations. However, Jess, with her unique understanding of how domestic violence disproportionately impacts women, recognized that our standard policies might not be adequate for protecting the personally identifiable information (PII) of the victims from their spouses.

Jess suggested that we adjust our recommended written privacy policies for this particular partner to account for the unique challenges they faced. Her insights stemmed from her knowledge about the specific risks that domestic violence survivors often encounter, which gave her a unique perspective that the male members of the team hadn't considered. By tailoring our approach to meet the specific needs of the non-profit, we were able to develop more robust security measures that better protected the sensitive information of the people they served.

Jess's contribution not only improved the security policies for the non-profit but also highlighted the importance of embracing diversity within a team. It would have been easy to become attached to something another team member wrote, to be defensive about someone suggesting an improvement (especially if that someone is different). However, by welcoming different perspectives, we can develop more comprehensive and effective solutions for our clients, ultimately leading to better outcomes and a more inclusive work environment.

Look at the team you are surrounded by. If they are all similar to you, you may not be in an ideal place to grow or evolve, to become better as a tech and rise to the top. Don't be afraid to suggest to a manager that you diversify your tech team. Look for organizations that embrace diversity in their organization and their tech teams. You will become a better technology professional because of it.

Part 7

Avoiding the Comfort Zone

Premise

If there was one thing Damien valued more than his bottle of Elijah Craig Toasted Barrel (a very tasty bourbon, if I say so myself), it was routine. The same black coffee every morning—two sugars and a dash of cream—a glance at the NASDAQ while he scoffed down a slice of buttered toast, and then off to work in his silver Volvo, which, like him, was a bit old school but damn reliable.

Now, don't get me wrong. Damien wasn't some boring stick-in-the-mud. He was one of those guys who'd quote Shakespeare in one breath and Seth McFarlane in the other. His cybersecurity consulting startup, Nightwatch, was his baby, and he ran it like a quirky sitcom dad - strict when it needed to be, but warm-hearted through and through.

On the other side of our tale is the fresh-out-of-college, always-on-her-toes, Megan. She was as eager as a puppy, but with the brains of a seasoned chess master. Hired by Damien for her vibrant energy and her degree in cybersecurity, she had just the right mix of curiosity and intellect to match the fast-paced world of technology. But she was a little more on the bleeding edge side of things.

Nightwatch was rolling down the hills of success partnering with LightWave Securities and Netshield Software: two trusted names in the business some time ago, but no longer on their A-game. Damien and his team knew these platforms like the back of their hands, after all they'd been using them for years. But, like the trusty old Volvo, they were reliable yet outdated. Their approach to cybersecurity was like playing Whack-A-Mole at the county fair. It was fun, sure, but not effective for the long haul.

It was just another day at the Nightwatch office. Damien sat at his desk, sipping his coffee and going over the monthly reports, while Megan walked in, headphones on, mouthing the lyrics to some Billie Eilish song. Little did they know, they were standing on the cusp of a whirlwind of change.

You see, in the world of cybersecurity, comfort is the silent enemy, like a mosquito that buzzes around your ear in the middle of the night. Sure, you can swat it away a couple of times, but sooner or later, you're going to have little choice but to get up and deal with it.

Now, Tuesdays had a notorious reputation at Nightwatch, known amongst the staff as 'Trouble Tuesdays.' This particular Tuesday was living up to its namesake. The phone rang on Megan's desk, its sharp trill echoing through her office. It was a frantic call from a major client, their voice ripe with panic. Their network had been breached, and a vicious malware was gnawing its way through their system.

"Look, Megan," they pleaded, "we trust Nightwatch, but we can't afford this kind of exposure. It's like we've been left out in the rain without an umbrella."

Megan was left staring at the slew of red alerts littering her screen like ominous Christmas lights. With her trusty - or rather, not so trusty - Netshield and LightWave tools, she felt like she was bringing a knife to a gunfight. The malware was spreading faster than she could halt it. Frustration bubbled inside her. She was left with little choice and had to completely cut off access to large chunks of the network in order to isolate the problems to buy her time.

Later that night, back at her home office, Megan scoured the depths of the internet for a solution. She had heard whispers of cutting-edge technology from companies like Asgard Security and Yellow Cloud Protection. They were said to be agile, intelligent and ruthlessly effective. It was time to see if the whispers held any truth.

Armed with free trials, Megan dove deep into Asgard Security and Yellow Cloud Protection's technical ecosystem reading docs, watching tutorials, and experimenting late into the night. Fueled by sheer determination (and too much caffeine), she unlocked their true potential: sophisticated, intuitive, and leagues ahead of what Nightwatch had been using.

She wanted to run and tell Damien and the rest of the Nightwatch team. She wanted to show them that there was a world beyond the comfort zone of LightWave and Netshield. But not yet. First, she needed to prove it could work.

With her newfound knowledge, Megan jumped back into the battleground. As she deftly navigated through cloud platform settings, she felt like a maestro conducting a symphony. With every click of her mouse and stroke of her keyboard, she wrestled control back from the malware.

It was a grueling battle, stretching into the early morning hours. But as the sun began to peak over the horizon, Megan finally saw it: a clean, green, malware-free alert. She had done it. She even went so far as to configure the endpoint detection and response settings, so there would be proactive alerting to the team in the future before an infection spread laterally across the network again. The first battle had been won. But the real challenge, convincing Nightwatch to step out of its comfort zone, was still to come.

As we take a step back to look at our story thus far, a couple of things become clear.

First, as in life and cybersecurity, a false sense of comfort is a disaster waiting to happen. In this case, it's Damien and his team at Nightwatch who've settled into the reassuring yet risky rhythm of the familiar. They're basking in the sun of success, too at ease to notice the storm clouds gathering. Why worry? They've got Netshield and LightWave in their arsenal. Never mind the fact these tools are as outdated as men's colonial wigs and makeup.

Then, we have our knight in shining armor, or should I say, our dame in shining boots—Megan. Fresh-faced, energetic, and unwilling to let comfort be her guide, she's like a shot of caffeine in Nightwatch's sleepy routine. When the 'Trouble Tuesday' hits, she doesn't freeze or give up, but she hits the ground running. Armed with just her determination and a couple of free trials from Asgard Security and Yellow Cloud Protection, she's off to learn new platforms, new technologies, and try to fix a real-time issue for a client, and perhaps a long term issue for Nightwatch.

Now, while I could take this opportunity to launch into a spirited rant about the benefits of continuous learning and certifications (notably absent in our friends at Nightwatch), I'll refrain for now. Instead, let's take a moment to appreciate the quiet heroism of Megan's actions. She didn't just stick a band-aid on the problem; she took it upon herself to find a more permanent solution.

But we're not quite out of the woods yet. The malware is at bay, and Megan's proven the worth of stepping out of the comfort zone. But can she convince the rest of the team, especially Damien, to follow suit? He might need to break into that bottle of Elijah Craig sooner than he thinks.

Survival is Insufficient

The next morning was one of those perfect, crisp days, the kind where the air is like the top of a creme brulee, slightly glazed yet promising the warmth of a new day. Megan walked into the Nightwatch office with a spring in her step, the exhilaration of the previous night still fresh in her mind.

Finding Damien in his office, nursing his morning coffee and mumbling something about the NASDAQ, she took a deep breath and knocked on the door. "Damien? Do you have a minute? We need to talk."

"Oh?" Damien raised a questioning eyebrow. "Did we run out of coffee filters again?"

"No. It's about LightWave and Netshield."

He choked on his coffee. "Sorry, what?"

"LightWave and Netshield," Megan repeated, "they're not cutting it."

"Excuse me?" Damien's voice was like a bucket of icy water.

Megan plowed on, explaining the malware attack and how she had single-handedly averted disaster using Asgard Security and Yellow Cloud Protection. But the more she spoke, the deeper Damien's frown became.

"You don't get it," Damien said, his voice tight. "Our clients trust us because we've been reliable. Because we don't just chase every shiny new thing that comes along."

Megan folded her arms. "Reliable? Damien, we almost lost a client yesterday. That's not reliability, that's stagnation."

The room was as tense as a stretched rubber band. Damien looked like he was about to explode, but then Megan said something that cut through the tension like a hot knife. "Damien, I get questions every day about the products we use. What we're using would be like telling customers our Internet Service Provider is

[155]

America Online and we use Netscape Navigator as our browser of choice. Today. In 2025. We might as well print MapQuest directions to MySpace HQ."

The words hung in the air. Damien's frown slowly dissolved into a thoughtful expression. He pursed his lips and looked at Megan. She was right. The world was changing. *It had changed.* And they had to change with it. Former giants in the industry weren't cutting it, new solutions and fresh ideas from cutting edge companies needed to be explored if Nightwatch wanted to be relevant and grow. One day you're relevant, the next you're not.

"You're right…" Damien slowly nodded. "Let's fix it. Tell me more about these new products you demo'd and used to fix the issue?"

Their conversation continued and Damien softened to the ideas Megan had brought to him. He was uncomfortable and nervous, but he recognized why this was an important discussion to have.

Over the next few weeks, Megan took the lead. The Nightwatch team ran Asgard Security and Yellow Cloud Protection side by side with LightWave and Netshield, testing the waters. There was resistance, of course. Even outright hostility in the beginning. But as the entire staff witnessed the undeniable superiority of the new platforms and saw the marked improvement in their clients' cybersecurity health, they slowly came around. Even the most stubborn of them had to admit that ticket volumes had significantly reduced, and their relationship with clients was better than ever.

It was hard. Learning new tools and processes always is. But with Megan spearheading the project, guiding and teaching along the way, they found their footing. Turns out, the new platforms weren't as daunting as they had thought. With time, they even came to appreciate the sleek user interfaces and intuitive features. They saw that these weren't just alternatives to their old platforms. They were significant upgrades. It just took some time, courage, and effort to realize it.

After two months, Nightwatch had completely migrated all their clients from LightWave and Netshield to Asgard Security and Yellow Cloud Protection. The move was met with positivity, even relief, from their clients, further solidifying the team's faith in their decision.

Damien stood in his office, looking over his team, hard at work on their shiny new platforms. He couldn't help but feel a pang of pride. And right in the middle of it all was Megan, guiding, teaching, and proving every day that pushing beyond the comfort zone was the only way to ensure growth.

One of the most important lessons I've learned in my career is that innovation waits for no one. The digital landscape changes faster than Elon Musk's fancies and we've got to adapt just as quickly if we want to keep up with the next trending Xeet on not Twitter. Like a minivan in a Formula 1 race, if you don't accelerate, you'll eat dust. That's a lesson Damien almost learned the hard way.

And before you start warming up those fingers to type up some hate-mail about how your company's firewall has been working fine since you updated it in anticipation of Y2K, I'm not just talking about updating your tools (though, seriously, replace that firewall, my dude). I'm talking about a mindset change. Cybersecurity isn't a stagnant field. It's not about being comfortable, it's about being prepared. It's about reading the fine print on those software agreements, and knowing that when it comes to cybersecurity, there's always a bigger, badder bug just waiting to take a bite.

What Megan understood, that took Damien a minute (or a major incident), was that cybersecurity is more than just protecting against threats; it's about anticipating them. The strength of a cybersecurity strategy lies in its ability to evolve, to keep up with the changing threat landscape. It's about pushing boundaries, and just as importantly, about pushing yourself.

If your mindset about change is as old as the outdated solutions you provide, you're doing yourself and your clients a disservice.

The real MVP of this tale isn't Asgard Security or Yellow Cloud Protection; it's Megan. She showed us that being fearless in the face of change, willing to learn and adapt, is what sets apart the average from the extraordinary. She's not just a beacon of hope for every young professional stepping into the field, but also a reminder to all the "Damiens" out there, that it's never too late to evolve.

And to the bosses who are still mulling over whether to switch from your trusty old platforms to something a bit more... modern, remember this: your comfort zone might feel like a cozy hammock on a sunny day, but in the world of technology, it's a ticking time bomb. So, keep learning, keep growing, and remember, the only constant in technology is change. The minute you feel too comfortable, that's when you should start worrying. Because comfort is the first sign of complacency. And we all know how that ends, don't we?

There's no harm in learning something new. Especially when that "something new" might just save your digital asse(t)s. Be like the Borg and adapt. Resistance is futile!

Embrace change...
- ➤ Change is not the enemy—stagnation is.
- ➤ View shifts in tech as opportunities rather than threats.

In my experience, embracing change has been essential for success. I recall a time when the company I worked for at the time was heavily focused on providing traditional on-premises infrastructure, such as file servers and Exchange servers, to our clients. While these solutions were familiar and comfortable to work with, I began to realize that they were not only more expensive for our clients but also limited our ability to grow and expand our skill set.

Years ago, one of the many clients I did work for, a medium-sized medical practice, needed a more efficient and cost-effective solution for their IT needs. As I began researching alternative technologies, I discovered the potential of Office 365, SharePoint, and Exchange Online. At first, I was hesitant to move away from the familiar on-premises solutions, fearing that the transition to cloud-based services might be too complex or risky. However, I recognized the importance of staying relevant in the industry and decided to embrace the change.

I took the initiative to learn about cloud technologies and their benefits, diving deep into the functionality and advantages of Office 365, SharePoint, and Exchange Online. As my understanding of these solutions grew, I realized that they could significantly improve the efficiency and cost-effectiveness of our clients' IT infrastructure, especially for the medical practice that we were assisting.

By embracing change and seeking out learning opportunities, I was able to transition our client to a more affordable and efficient cloud-based solution that met their evolving needs. This not only helped the medical practice save on IT costs but also enabled them to streamline their operations and focus on providing excellent patient care. In the process, I expanded my skill set, stayed ahead in the industry, and grew professionally. Embracing change ultimately allowed me to offer innovative solutions and better serve our clients.

Stay informed: It's essential to keep up-to-date with industry trends, emerging technologies, and best practices.

By staying informed, you'll be prepared to adapt your approaches and strategies to stay relevant in a constantly changing landscape. This requires research, paying attention to the news, the industry, reading and watching relevant videos, pursuing certifications and education where possible. It could even be as simple as demoing different products, even if you're happy with the ones you know and use already, much like Megan did in our fable.

[162]

Be open to new ideas: Being receptive to new approaches, even if they challenge your existing beliefs or methods, is essential for adaptability.

We saw how not being adaptable almost bit Damien in the butt. Here's another actual story from my career. There was a time when a colleague named Frank suggested offering "firewall as a service" to our customers. This idea was entirely new to me and the rest of our team, as we were used to configuring firewalls as part of larger capex (capital expenditure) projects and having our clients purchase the hardware and renew licensing on their own each year.

We decided to keep an open mind and explore the benefits of providing firewall as a service. Frank's idea was to buy the firewall devices, configure them according to our best practices and standards, and then rent the firewalls to our customers for a monthly fee. We would own the hardware and be entirely responsible for licensing and managing the firewall, while the customer would sign a 36-month contract for the service.

As we delved deeper into the concept, we quickly realized that Frank's idea could create a win-win situation for both our clients and our company. By bundling all the costs and offering the service on an opex (operational expenditure) monthly basis, we made it more affordable for our clients, while still ensuring profitability for us over the life of the contract.

We decided to give Frank's idea a try, and it turned out to be an incredible success. Providing firewall as a service became a new product offering for our company, generating significant income and expanding our portfolio of services. By being open to new ideas and willing to explore alternative solutions, we were able to better serve our clients and stay at the forefront of the industry.

If you want to be a great tech professional, you need to think outside the box and embrace ideas that are not the norm. Be a part of that process, both as someone who thinks of these types of ideas and as someone who listens to others that have those ideas.

Learn from mistakes: Mistakes are inevitable, but they also offer valuable opportunities to learn, adapt, and improve.

More recently, I had the opportunity to learn from a fellow project engineer named Stephen, who was tasked with migrating a client from a GoDaddy Office 365 tenant to a standard Microsoft Office 365 tenant. Office 365 data was not backed up by default, and additional licensing, retention policies, or third-party backup solutions were necessary to ensure data preservation.

Stephen diligently backed up all the mailbox data, but unfortunately, he overlooked the user OneDrive data. The migration process went smoothly, and everything appeared to be in order. It wasn't until after the old GoDaddy tenant had been deleted that we realized the mistake. Even though GoDaddy was supposed to be backing up the data, they had also deleted the backup archives when the tenant was deleted. There's nothing worse than that gut-wrenching knot in the pit of your stomach when you realize that you've lost data.

For several weeks, the entire team worked tirelessly, collaborating with vendors and exploring every possible avenue to locate and restore the missing data. Despite our best efforts, the data was irrevocable. This experience was a stark reminder of the importance of thorough planning and double-checking every aspect of a project.

From this incident, we learned valuable lessons about data management and backup processes. We began implementing more rigorous checks and balances to ensure all data was accounted for and properly backed up during future migrations. Additionally, we emphasized the importance of learning from mistakes and using setbacks as opportunities for growth and improvement. By doing so, we became more adaptable and better equipped to handle future challenges.

Cultivating adaptability to avoid the comfort zone is an essential key to thriving in the dynamic IT industry. By embracing change, staying informed, being open to new ideas, and learning from mistakes, you'll be better prepared to navigate the ever-evolving landscape and provide innovative solutions to your clients or employer.

Part 8

Problem Solving and Creativity

Logic Games

The most fundamental part of a career in information technology, cybersecurity, or programming is your critical thinking ability: the underlying logic to your thought process, reasoning, and problem solving.

Technical troubleshooting is the application of critical thinking to technical problems. But the core ability to analyze complex situations, identify underlying issues, and develop creative solutions isn't inherently technical.

Understanding and honing that will not only set you apart as an exceptional IT professional, but it will supercharge your growth and give you many options for advancement. The best professionals aren't just problem solvers, they're adaptable thinkers who anticipate issues before they happen. The difference between a great IT tech and a future CTO is the ability to think critically beyond just fixing problems.

Riddle me this, reader...

A multinational corporation is moving its data center to a new location. They have five large databases named after Roman gods: Mars, Venus, Neptune, Apollo, and Diana, which need to be transferred over the network to the new location.

The databases have different sizes and importance to the company:

- The Mars database is the largest and is twice the size of the Venus database. Each hour of Mars downtime costs the company $10,000.
- The Venus and Apollo databases are of the same size. Each hour of Venus downtime costs the company $8,000. Apollo is the least critical database and its downtime costs the company $2,000 per hour.
- The Neptune database is smaller than Mars and Venus, it's exactly half the size of the Mars database. Each hour of Neptune downtime costs the company $3,000.
- The Diana database is the smallest and is exactly half the size of the Apollo database. Each hour of Diana downtime costs the company $6,000.

Due to network limitations, the company can only transfer one database at a time and larger databases take more time to transfer- the transfer rate is such that a database's size directly translates to transfer time in hours.

During the transfer of a database, it is unavailable for use, which we'll refer to as "downtime". The corporation aims to minimize the total cost due to downtime.

Given these circumstances, what is the most cost-effective order to transfer the databases?

- **A.** Mars, Venus, Neptune, Apollo, Diana
- **B.** Diana, Venus, Mars, Apollo, Neptune
- **C.** Mars, Neptune, Venus, Diana, Apollo
- **D.** Venus, Diana, Mars, Neptune, Apollo
- **E.** Diana, Venus, Mars, Neptune, Apollo

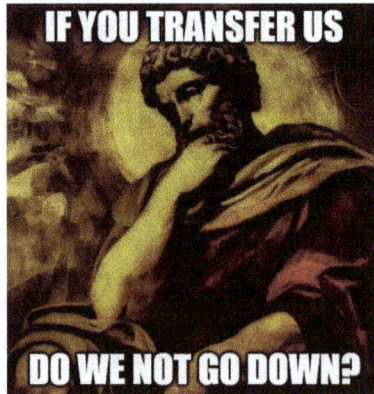

Go ahead, take your time. I've got another one waiting for you!

Cue Final Jeopardy Theme Music

How about a Network Outage Investigation?

A network technician is troubleshooting a sudden network disruption at a company. The company's digital infrastructure consists of four servers named after Greek gods: Zeus, Poseidon, Athena, and Hermes. The servers, based either on-premises or in the cloud, support different functions for the company's employees.

Here are the details the technician knows:

- Zeus is an on-premises server. When Zeus is functioning correctly, it supports 80 onsite employees and also ensures that Athena, a cloud server, remains operational.
- Athena is a cloud server dependent on Zeus's operation. It supports 30 remote employees.
- Hermes is a cloud server. When Hermes experiences problems, it can cause disruptions in Zeus's operation while leaving Poseidon unaffected. Hermes supports 60 remote employees.
- Poseidon is an on-premises server. It operates independently of the others and supports 80 onsite employees.

Currently, Zeus and Athena are not functioning, whereas Poseidon and Hermes appear to be operational. The technician needs to determine the most likely cause of Zeus and Athena's outage and restore functionality for the maximum number of employees in the shortest possible time.

Given these details, which server should the technician prioritize for inspection?

A. Zeus

B. Athena

C. Poseidon

D. Hermes

E. Inspect Zeus and Athena simultaneously

YU NO GIVE US ANSWER?!

TEACH US TO BRAIN GUD!

Alright, alright, how about something a bit more fun?

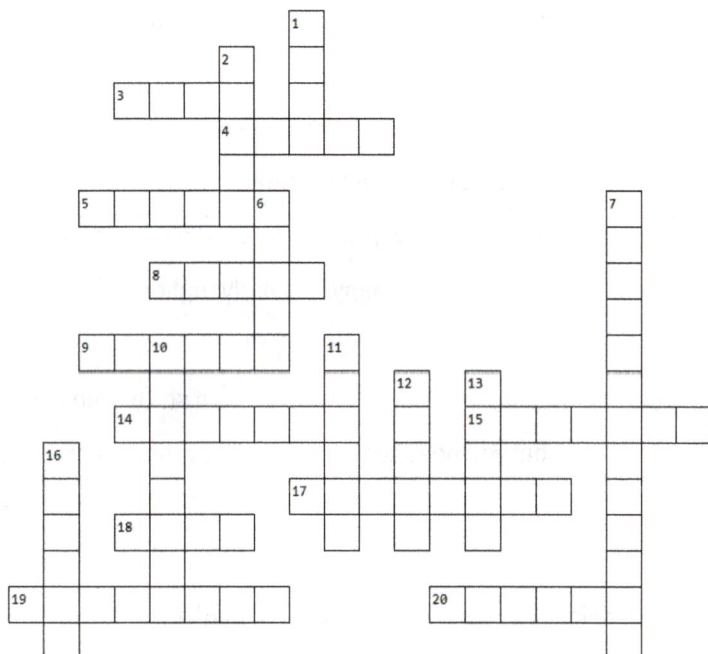

Across

3. Acronym for digital talking
4. Pirates have them (usually the eye variety)
5. Someone with a machete (or a remote access toolkit)
8. What exterminators do when they tent your house
9. 365 times better than Google Workspace
14. Sometimes you just need to take a baseball bat to it...
15. Nathan Fillion's Browncoat (Not quite)
17. Type of blanket
18. Lieutenant Commander Android on the USS Enterprise
19. Hundreds of years ago, all homes were (blank)
20. The first step of all troubleshooting...

Down

1. Pat Sajak or Jadzia Dax
2. If you eat it, you'll realize you're naked
6. Paperboys have them
7. Kryptonite is Superman's greatest (blank)
10. Sometimes required for construction to pass inspection, sometimes it does its own inspecting!
11. Don't be rude, tip them at least 18%
12. Japanese pancakes
13. The original way to slide into someone's DMs
16. A realm of administrative autonomy

These logic games are not meant to trick you, but they are excellent ways to help you gauge your own reasoning skills.

The first question's answer is E. This answer optimizes for size-to-cost ratio in terms of downtime for each database. Option A is a trick answer, as this only prioritizes based on database size and not the cost for the company. Similarly, option B is almost correct, but not quite the most efficient. Option C is fairly easy to rule out, and option D may seem appealing as it starts with a high cost database, but you need to think about the ratio of downtime to cost.

As a technical person, you're likely inclined to think only of technical impact, not overall business impact. But to elevate your career, you need to consider business impact. That type of foresight and thoughtfulness will lead to promotions and other career opportunities in management, directorship, and even C-level roles at companies.

For the second question the correct answer is option D. Hermes is critical to the functionality of Zeus, which in turn is critical to the functionality of Athena. By looking at Hermes first, you're trying to kill 2 birds with one stone. It's more likely a single issue causing the problem than 2 isolated incidents (thanks, Occam!).

The trick answer here is E. It's enticing to focus on the two down servers simultaneously, but it won't be fruitful if the cause of Athena being down is Zeus being down, and Zeus being down is due to a dependency on Hermes functioning properly. Option A is also a bit of a trick. It's more focused than Option E, but it still ignores Zeus' dependency on Hermes to function properly. Options B and C are a bit easier to rule out from the get-go.

For the crossword puzzle, how about I give you the answers to each question, but I encrypt the answers first? You can get them by decrypting them (hint – it's a Caesar cipher of 3).

Across

3. yrls
4. sdwfk
5. kdfnhu
8. ghexj
9. riilfh
14. sulqwhu
15. pdozduh
17. vhfxulwb
18. gdwd
19. zluhohvv
20. uherrw

Down

1. krvw
2. dssoh
6. urxwh
7. yxoqhudelolwb
10. iluhzdoo
11. vhuyhu
12. forxg
13. hpdlo
16. grpdlq

If you were able to solve these questions and the crossword puzzle with minimal help or no help at all, then you've probably got some pretty decent underlying critical thinking ability. Importantly, more than just having those abilities, you'll want to hone them.

Practice analyzing situations from multiple perspectives, questioning assumptions, and weighing pros and cons.

Honing your critical thinking skills is essential for problem-solving. It enables you to assess a situation objectively, identify potential issues, and evaluate different solutions. By considering various perspectives and challenging assumptions, you can make more informed decisions and find innovative solutions to complex problems. This is the very basis for technical troubleshooting; however, when you focus too heavily on technical problems and technical solutions, you may not be truly thinking critically. Here are a couple of examples.

In one instance, I was working with an organization that was notorious for getting their computers infected with malware and ransomware. We initially implemented an email filtering solution, but that didn't stop them from browsing the Internet and clicking malicious links. Next, we instituted DNS protection and content filters, but users would plug in USB drives from home with files they downloaded on their home computers. Sometimes the malware got past their default endpoint protection software, so we would have to download and use a second-opinion scanner like Malwarebytes to clean it up. This went on for months.

We kept trying to use technology to solve the malware infection problem, leading to months of protracted downtime and wasted technical troubleshooting by multiple technicians. Eventually, we realized that the issue boiled down to security awareness training and education. By suggesting and providing user security awareness training, we alleviated the pain we were all experiencing and greatly reduced the incidents of malware on the network. This experience taught me that even if a problem seems technical at first glance, not every issue an organization faces will be technical or require an overly technical solution. Critical thinking should let you look beyond the technical and consider aspects like operational issues.

In another example, our team was faced with a client who was constantly experiencing communication breakdowns and project delays. While our initial instinct was to provide them with a more advanced project management solution, we took a step back and analyzed the situation more critically. Upon closer examination, we discovered that the root cause was not the technology but rather a lack of clear roles and responsibilities within the client's team. Instead of implementing a new technical solution, we worked with the client to develop a more effective process and structure for them, including clearly defined roles, responsibilities, and communication channels.

This non-technical approach addressed the underlying issue and led to improved collaboration and efficiency within the client's team. It also reinforced the importance of non-technical problem-solving skills in my career. Developing these types of critical thinking skills is a process that can be enhanced through various methods and practices.

Ask questions.

Cultivate a curious mindset and consistently ask questions about the information and ideas you encounter. This will help you analyze situations more deeply and challenge assumptions.

Evaluate evidence.

Examine the evidence supporting various claims or arguments, and assess its credibility, relevance, and accuracy. This will help you make well-informed decisions and judgments.

Reflect on your own thinking process.

Periodically review and analyze your thought processes to identify areas for improvement. Consider whether you're relying too heavily on certain assumptions or biases and seek to adjust your thinking accordingly.

Practice problem-solving.

Engage in exercises and activities that challenge your problem-solving skills, such as puzzles, brain teasers, or strategic games (like the ones I created above). This can help sharpen your critical thinking abilities and teach you to approach problems from various angles. Crossword puzzles and memory games can help with this. If you want to go really hardcore, look into questions for the Law School Admissions Test (LSAT).

Read widely.

Expose yourself to diverse topics, authors, and perspectives through reading. This can help you gain a broader understanding of the world and develop your ability to think critically about various subjects.

Participate in discussions and debates.

Engage in conversations and debates with others on various topics. This can help you refine your arguments, learn to consider opposing viewpoints, and develop your critical thinking skills.

Seek feedback.

Ask for feedback from others about your thought processes and problem-solving abilities. This can help you identify areas for improvement and enhance your critical thinking skills.

By practicing these methods and incorporating them into your daily life, you can gradually enhance your critical thinking skills and become a more effective problem solver. This will pay dividends in your technology career, paving the way for senior roles, management roles, and differentiating yourself from the rest of the pack.

Cultivating Creativity

While logical reasoning and critical thinking are supremely important to a career in technology, it's just as paramount to embrace creativity and thinking outside the box. Don't fall into the trap of thinking there's only one way of doing something. Foster an open-minded, curious mentality that is willing to explore unconventional solutions.

Creativity and open-mindedness may not come naturally to everyone, though. At risk of sounding like a self-help book or a mom blog, I will give you some examples of creativity and also some ideas for how to become more open minded and develop your own creativity a little more as well.

Brainstorming sessions.

Regularly engage in brainstorming sessions, either individually or with a group. This encourages the free flow of ideas and helps you come up with innovative solutions to problems. **"But how do I brainstorm with myself?"** I'm glad you asked! Talk to yourself. Seriously. Compartmentalize one version of yourself with another. Become your own Devil's advocate! Some people choose to practice in the mirror, which is a method you could adopt as well. After you get over the initial "silliness" of talking to yourself, it really can be a very useful tool.

Cross-disciplinary learning

Learn about subjects and fields outside of your primary area of expertise. This can help you make unexpected connections and discover innovative solutions that may not be obvious within your own field. It may sound strange, but experimenting as a home chef, different cooking techniques, flavors, textures, etc. has made me a better technology professional (I can use cooking analogies for conveying complex concepts).

Try something new you think you'll be bad at or won't like.

We've previously discussed how recognizing failure is an opportunity for growth and improvement. Treating setbacks as valuable learning experiences and using them to iterate and improve your ideas is crucial. But let's take it a step further. Try something new that you think you'll be no good at doing. Afraid you have no rhythm? Take dancing lessons. Oh, you hated fish sticks as a kid so now you won't try shellfish? Get that lobster roll. If you don't like it, who cares? At least you tried it. Adopt a mentality of "Sure, I'll try it" no matter how afraid or sure you are about it. Not only will you probably surprise yourself, but this will also literally transform your way of thinking and allow you to be more open-minded in your career as well.

Set Constraints

Think of constraints as your creativity's personal trainer – they push your boundaries and force you to flex those innovative muscles. Set up a challenge for yourself, like solving a problem using the least amount of resources, or creating a solution within a very tight deadline. It's like making a gourmet meal with only five ingredients; not only does it get the job done, but it also forces you to be resourceful and inventive.

Use Analogies and Metaphors

You know how a caterpillar transforms into a butterfly? That's your idea, and you're the caterpillar. Analogies and metaphors can metamorphose a mundane concept into something spectacular. They're not just the spices in the dish of communication; they're the secret sauce that makes complex ideas digestible.

Engage in Creative Exercises

Unleash your inner Da Vinci or Shakespeare. Doodle, scribble, compose a haiku about cloud computing. Engaging in seemingly unrelated creative activities can surprisingly jumpstart your tech brain. Think of it as cross-training: you're not just a techie, you're an artist too.

Take Breaks and Change Your Environment

Ever notice how great ideas often hit you in the shower? That's no coincidence. Stepping away from the screen and changing your environment can reboot your brain. It's like giving your mind a mini-vacation, allowing it to return refreshed and brimming with new ideas.

Practice Mindfulness

Mindfulness isn't just a buzzword; it's about being present in the moment. It's like defragmenting your hard drive. Meditation, yoga, or just a few minutes of deep breathing can clear the mental clutter, making room for innovative thoughts and ideas.

Break Problems Down

Tackle complex problems like you'd eat an elephant – one bite at a time. Breaking them down into smaller pieces makes them less daunting and easier to manage. It's like solving a jigsaw puzzle; piece by piece, the bigger picture becomes clear.

Learn from Others

Don't reinvent the wheel – sometimes the best ideas are already out there. Look outside your field. Study a famous architect's approach to design, or a composer's method in music. It's like being a tech detective; gather clues from different sources and piece them together to solve your IT mysteries.

By incorporating these techniques into your daily life, you can foster ingenuity and innovation, ultimately in service to the goal of being a better technology professional. But, if you're looking for something a bit more personal and less generic than blog-like recommendations, here's how I incorporate logical reasoning and creativity into my life:

My evenings are usually a bit unconventional. Instead of the typical unwind in front of the TV watching something mindless, it's more like a mini-brain Olympics. The backdrop is Jeopardy, where my significant other and I vie to outdo each other with rapid-fire answers. Ken Jenning's voice is practically the soundtrack of our evenings. After Jeopardy is over, we'll turn to other game shows. Even Family Feud can be good to think outside the box. And the night's not just about shouting answers at the screen; there's a whole lineup of brain teasers waiting for us. Wordle, Spelling Bee, Connections, and a daily crossword are all part of our mental gymnastics (thanks, New York Times). There are some other fun and clever word games we often get absorbed into as well. It's our way of keeping the neurons firing. But this helps keep me sharp and motivated.

But why stop at word puzzles? Enter the world of tabletop gaming. Imagine a bunch of tweens (this is an old story for me) diving headfirst into the roles of cunning, albeit slightly inept, goblins. One of our quests? To solve a riddle that seemed straightforward but turned into a comedy of errors (hey, we were young). "I speak without a mouth and hear without ears. I have no body, but I come alive with wind. What am I?" Initially, our guesses were so hilariously bad or juvenile that we might as well have been in a Monty Python sketch. But amidst the laughter

[185]

and wild speculations, we eventually stumbled upon the answer (a few good dice rolls didn't hurt either): "an echo." This journey from bewildered goblins to triumphant riddle-solvers was a milestone moment for us younglings that the path to enlightenment (or at least to the next level of the dungeon) is filled with wrong turns, laughter, and learning. Since then, I've played in and game mastered dozens if not hundreds of other tabletop and live action roleplaying games. That experience has been invaluable in my professional growth and mindset.

This blend of Jeopardy, puzzles, and roleplaying adventures isn't just about having a good time. It's a gym for your brain, where logic and creativity get to flex their muscles. In the technical world, where problems can be as complex as a gnarly piece of code or as nuanced as a single user experience issue, this kind of mental agility is invaluable. It's about being able to pivot quickly, think on your feet, and approach challenges from multiple angles.

Bringing this mindset to your work isn't just beneficial; it's essential. The tech landscape is constantly evolving, and staying sharp is key to not just keeping up but leading the charge. So, consider this an invitation to mix a little play into your professional development. Dive into puzzles, challenge yourself and others, and remember that every misstep is just a steppingstone to a solution.

And if you're ever doubting the value of this approach, just remember the goblin crew and their riddle. It's a testament to the power of perseverance, creativity, and the ability to laugh at ourselves when we miss the mark. So, embrace your inner nerd, keep Jeopardy turned on in the background, and let the world of puzzles and games enrich your professional journey. After all, the skills you develop might just be the ace up your sleeve in your next big project or job interview. Plus, you'll have some great stories to share. I mean who doesn't love a good "we thought the answer was a fart" story at the water cooler?

Challenge yourself: Over the next week, pick one problem at work or in daily life and apply a new problem-solving method to it. Break it down, approach it from a different angle, or use a technique from this chapter. The best way to sharpen your skills is to use them. So start now.

Section X (Definitely Not Part 9)

Branding and Peopling

Perception is Reality

Do you know why Robert Downey Jr.'s casting for Tony Stark in the Marvel Cinematic Universe is considered so iconic? I mean, sure, he is an incredible actor. Prior to his Iron Man casting his career wasn't doing so well. But with Iron Man, as Tony Stark, he rebranded himself. He completely reinvented his image. In many ways he became synonymous with the character. He's beloved by fans and respected by peers now more than ever.

Now, if RDJ was a terrible actor, he may not have been able to rebrand himself. So, his acting ability certainly had a hand in his success. But it wasn't just his acting, it was his ability (and Marvel's) to capitalize on that perception and emphasize it. And it had a sort of positive-feedback-loop effect.

What about Taylor Swift? Is she the best vocalist and songwriter of all time? I don't think she would say so herself… (to be clear, I am not coming at Taylor Swift, I think she's an incredibly talented and intelligent woman with great business acumen). Talented she may be, the most talented debatable, but one thing is for certain: she's one of the most successful and most world-recognized figures of all time. Her music is hers; her image is hers, and everyone knows it.

Now, picture your celebrity crush in their prime. I know you've got that mental image of an iconic photograph or a poster that was hanging up in your bedroom when you were 16 years old. I'd bet you $100 that photo or poster is airbrushed and filtered. Does it matter, though? Does it make your perception of their attractiveness any less real?

Every day, people are doing things to build a brand. It is a relatively recent phenomenon for many individuals to do this, but it's the way of the world now. Influencers, business moguls, pop stars, and yes, everyday professionals like me and like you, are (or should be) curating an intentional, thoughtful, and accurate (if not polished) brand for ourselves.

Take the photo and title I chose for this Part 9... uhm I mean Section X... of the book. Despite what may look like a case of obvious narcissism (yes, that is a semi-transparent photo of yours truly on top of a city scape background), the truth is, that photo is an AI-doctored photo. Imperfections have been whisked away, my weight has been reduced, a flattering angle and filter has been utilized, and I even used a transparency technique to set it on top of a much brighter looking cityscape. It's a photo I've used as a profile picture on social media (sans city scape). Why? Because I, like most people on social media, want to project a specific image in the hopes that other people see me how I want to be seen (maybe even how I want to truly be). It's a rudimentary type of branding.

What about the title change? It's a rebrand. I did not call it a "Part" I called it a "Section." I didn't use the next number in line, I used the letter X (also Roman numeral for 10). Kind of like the jump from Windows 8 to Windows 10, skipping over 9 completely (don't start on about Windows 8.1) or Windows 98 to XP (let's pretend Windows ME never happened and 2000 was really only marketed to businesses). Did it have an effect? Did it catch your attention, either in a good way or a bad way?

Now, I am not a branding or marketing specialist. But I understand the basics and the value of projecting an image (literally and figuratively). Branding is beyond just something physical or tangible like my photo or section title. Take my other examples, you absolutely need to have skills (or looks in some cases) to backup whatever brand you're building for yourself. That's like the difference between a really great, enhanced photo of yourself and catfishing on a dating website as someone completely different. Curating a professional brand is no different. You need the skills, knowledge, experience, talent, whatever to back up the brand you're trying to build for yourself. You don't have to be the best, but you have to be good. Everything else is window dressing, presentation, showing off that brand and convincing people why your brand is the right brand to choose.

In the context of professional branding in the tech world, this could be as simple as creating and perpetuating a specific reputation where you currently work, or as in-depth as having a YouTube Channel, one or more blogs, a personal website, books, a stellar LinkedIn page, and more. It all sort of depends on a combination of your motivation, comfort level, and goals. The less far-reaching and more subtle your brand is, the lower the impact it will have on you and your career.

Perception is reality.

Just like how clients or end users will perceive things in a specific way, so will colleagues, bosses, and prospective employers.

But, Like I mentioned earlier, before you start to build a personal, professional brand for yourself, you need the skills or knowledge to back it up. So that's where we're going to start.

Continuous Learning and Networking

This industry evolves at breakneck speed—a new programming language, new cybersecurity threat, new emerging technology, and more pops up before you've even mastered the last one. But here's the thing: this relentless pace of change isn't just a challenge; it's an opportunity. An opportunity for what, you ask? For continuous learning, adaptation, and, most importantly, for personal growth.

Now, I know we've touched on this previously, but we're going to dial it up a notch. In an ocean of change, some folks are content floating along in their inflatable donuts, clutching their first and only bits of tech knowledge like a security blanket. But not you. You're not just here to float. You're here to sail, to explore, to conquer new landscapes in technology and in your career. So how do you do that? By embracing change as if it were a long-lost friend.

Continuous learning in tech is like keeping yourself properly updated and patched; fail to do it, and you risk becoming obsolete or a weak spot, a risk to keep at an organization. Remember, the goal isn't to know everything. That's impossible and, frankly, exhausting. The goal is to stay curious, to keep asking questions, and to keep pushing the boundaries of what you know. Whether it's picking up a new programming language, dabbling in new cloud technologies, or venturing into penetration testing, every new skill you acquire is a new tool in your arsenal, a new way to view and solve problems. It broadens your skills, makes you more valuable, and will allow you to command higher salaries.

But let's take a step back from the technical for a moment. As I've demonstrated throughout most of our stories, personal growth in tech isn't just about the hard skills; it's about the soft ones too. Communication, emotional intelligence, teamwork— these are the skills that turn a good technologist into a great one. After all, what's the use of writing elegant code or doing comprehensive solutions architecture if you can't explain it to your team, if you can't empathize with the end-users, or if you can't collaborate to bring a project across the finish line?

The tech industry, at its heart, is about solving problems for people. And to do that effectively, we need to understand them, connect with them, work with them. It's not enough to be brilliant behind a screen. You need to be brilliant in the boardroom, in the team meeting, in the interview, with a new prospect. This dual focus on technical mastery and interpersonal skills is what will eventually make your personal brand not just recognizable but revered, ultimately leading you to professional success.

As you navigate the waters of this industry, remember: the waves of change are inevitable, but they're also invigorating. They push us to learn, to grow, to evolve. They challenge us to not only keep up with the industry but to lead the way in it. And in doing so, they offer us the chance to redefine not just our careers, but ourselves.

So, are you on board with the whole "embracing change" and "continuous learning" spiel? Fantastic! Here's some lemon and Tabasco for your oysters.

First up: books. If you're reading this one, you should be open to reading others. Tech books can be an invaluable resource for learning and growth. They range from the foundational (think "Clean Code" by Robert C. Martin) to the niche and provocative (ever heard of "The Art of Deception" by Kevin Mitnick for social engineering insights?). They don't have to be technical either. I previously mentioned "The 7 Habits of Highly Effective People" by Stephen Covey. Books are your deep dives, your long-term relationships in the fast-paced world of your career. And the best part? There's a book out there for virtually every topic under the digital sun. Use Google, find ones that look interesting or informative, and then read them, absorb them, and use them to get better.

Your digital diet should include more than just books, though. Add in a healthy serving of blogs and podcasts. They are the snackable content that keeps you informed about the latest threats, technologies, and industry trends. Sites like Hacker News, TechCrunch, or individual thought leaders' blogs offer insights, trends, and the occasional tech gossip. Podcasts, meanwhile, let you absorb knowledge while you're commuting, working out, or just chilling. "Security Now" is a fantastic source that blends security expertise with accessibility. They keep you abreast of the latest threats, trends, and tales from the front lines.

However, just staying apprised of trends and the latest info isn't good enough to keep your skills relevant. What you learn by engaging in these resources is more akin to staying well-rounded and thoughtful in a generic sense. But we're not looking for generic. That's where certifications come into play.

Certifications are both your banner and your armor. They declare your expertise and your allegiance to the ongoing quest for knowledge. Whether it's the strategic oversight provided by CISM, the foundational fortitude of various Microsoft cloud certifications, or the network nous of Cisco's CCNA, each certification is a milestone on your professional journey, a testament to your commitment and competence. Certifications alone are enough to garner interest from prospective employers and add clout to your reputation. Combine them with everything else I've tried to impart, and you will have a behemoth foundation upon which to build your brand.

But getting certifications is easier said than done. In my experience, the easiest way to get certifications is through a combination of hands-on experience (the actual jobs you're working) and training resources designed to get you those certs. Platforms like Pluralsight, Coursera, Udemy, and Cybrary are treasure troves of technical knowledge and training. "For Dummies" books are also great practical resources for beginners. From bootcamps to courses on AI, blockchain, programming, or even ethical hacking, these resources offer structured learning paths for every skill level. Even brain dumps can be useful for getting certs (though don't rely on those... remember you want to have the skills to back it up, and a brain dump may get you the cert, but it won't give you the knowledge). With all of these at your disposal you can easily snag these certifications to jazz up your LinkedIn profile and resumes while retaining the knowledge necessary to get them.

One important reminder: as you acquire these skills and certifications, you should focus on understanding the principles that underpin these topics. Getting the CCNA doesn't give you the excuse to say "I know Cisco firewalls and networking, but I don't know Fortinet or SonicWall." While the CCNA may be more focused on Cisco, the foundation and principles you're learning can be applied across vendors. At that point it becomes much more about becoming intimately familiar an interface than it does understanding the core technical concepts. This is an extremely important lesson that you should take to heart in order to separate yourself technically from your peers.

So, what's next? You've adopted a mindset that embraces change. You're now exercising your mind and creativity. You are committed to the idea of solving challenging technical problems without foregoing important soft skills. You know you need to build a brand for yourself. You're staying apprised of trends and remaining knowledgeable about contemporary industry-specific needs and challenges. You've got some certifications under your belt and have strong foundational knowledge (not just a vendor-specific skillset attitude). You're working a job somewhere and you are itching to do even more to grow your career. Well, you may have heard a cliché before: it's not just about what you know or what you can do, it's about who you know. This is where networking and workshops come into play.

I know many of you may be diehard introverts, but unfortunately one of the best paths to building a brand and growing your career is through networking and getting to know other professionals and technology organizations. Once the world of books, blogs, and online courses has filled your brain to the brim, it's time to step out into the wild. Conferences and meetups are fantastic for learning cutting-edge techniques, hearing from industry leaders, and, yes, networking. They're where the digital meets the physical in a symphony of shared knowledge and geekery. Events like DEF CON and IT Nation are not just gatherings; they are pilgrimages for those of us seeking the cutting edge of our craft. The knowledge shared, the hands-on sessions, the serendipitous encounters all weave into the rich tapestry of our professional lives. You'll meet other folks just like you, you can share war stories, see cool new gadgets, learn about various platforms, and start engaging in a multitude of tools and demos to expand your hands-on skills. They also give you more insight into what other professionals are doing. Not only does this offer the opportunity to grow your skills and knowledge, but you will also be establishing connections, building rapport with companies and potential colleagues. This is paramount to building a brand and setting yourself up for career success.

Last, but certainly not least, don't forget about other people you already know or work with. Colleagues, mentorships, study groups, online forums—learning from and with others is perhaps the most dynamic way to grow, learn, and establish contacts. The tech community is vast and varied, with veterans and newcomers alike contributing to the conversation. In the end, our journey is one we share with fellow travelers. The networks and communities we build and participate in are the lifeblood of our profession. GitHub, Reddit, LinkedIn, even forums like Spiceworks can be goldmines of collaboration and insight. They're not just platforms; they're virtual roundtables where we come together to share, solve, and sometimes, simply to stand in solidarity. They remind us that, though our challenges may be complex, we do not face them alone. And the more you participate and become familiar, the more you outreach, the more you will learn and build on a personal contact list of potential references, cheerleaders, and employers.

Overcoming Imposter Syndrome

Before we circle back, I want to talk about impostor syndrome: the unwelcome guest in the minds of nearly everyone, from the greenest of novices to the most seasoned experts. It's that nagging voice that whispers, "You're not good enough," "You don't belong here," or "Sooner or later, they'll figure out you're a fraud." Sound familiar? You're not alone. But here's the good news: feeling like an impostor is often a sign that you're growing, that you're pushing yourself into new, uncomfortable territories. And that's exactly where you need to be.

First off, let's debunk a myth. Impostor syndrome does not equate to a lack of skill or knowledge. Quite the opposite. It tends to prey on the highly skilled and ambitious. It's a distortion of self-perception, not a reflection of reality. So, how do we fight this phantom?

First, acknowledge and name It. Recognizing when impostor feelings are at play is half the battle. By simply acknowledging these feelings without judgment, you're taking the first step towards disarming them. Don't stop at this, though. One of the most powerful antidotes to impostor syndrome is conversation. Talk to your mentors, peers, or even a professional if it's seriously impacting your life. You'll be surprised to find how common these feelings are and how sharing them can lighten the load. You should also make it a habit to reflect on and celebrate your achievements, no matter how small they may seem. Completed a project? Learned a new skill? Took the initiative on something at work? Celebrate it. These are tangible proofs of your competence and growth. In tech, the goalposts are always moving. Instead of aiming for perfection, aim for continuous improvement. Each mistake is a learning opportunity, not a mark of failure. And last, regularly seek feedback, not as a validation of your worst fears, but as a roadmap for growth. Constructive criticism can help you understand your areas of improvement and highlight your strengths in ways you might not see yourself.

Throughout my career, spanning the realms of systems administration, networking, cybersecurity, and beyond, I've had my own fair share of encounters with impostor syndrome. Despite accolades, successful projects, and the respect of my peers, there were times I found myself questioning my own abilities. "Do I truly belong here?" "Am I good enough?" "Am I really qualified to do this job?" These questions, and others like them, have been unwelcome companions on my professional journey.

What I've come to realize, though, is that impostor syndrome never truly goes away. It's like the background noise of a busy street; you might get used to it and ignore it usually, but every so often, it demands your attention. The key isn't to silence it entirely (a Herculean task that's perhaps more fantasy than reality) but to learn how to manage it, to turn that noise into music, or at least into something you can work with.

For me, the most effective weapon against impostor syndrome has been subject matter expertise. And the measure for if I considered myself a subject matter expert has been whether or not I could teach it to someone else with confidence and clarity. There's something about the process of learning something so thoroughly that it moves from the external (something you know) to the internal (part of who you are). And there's an added layer of validation that comes from teaching that knowledge to others, from seeing the light of understanding flicker on in someone else's eyes. If others are relying on you to explain something to them in a way they can understand, then you've made it. It's proof, irrefutable proof, that you know your stuff.

This approach has served me in two crucial ways. First, it forces me to confront my areas of insecurity head-on. If I'm feeling unsure about my expertise in a new cybersecurity threat or a complex networking concept, I dive deep. I research, I study, I experiment until I feel like I could write a book on the subject. Then, I share that knowledge through mentorship, workshops, writing, training videos, meetings, or any platform available. Teaching is the test, and it's one you can't fake your way through.

Second, it reaffirms my value, not just to myself, but to my peers and the industry at large. Every time I successfully teach a concept, it's a reminder that I have something valuable to offer, that my place in this industry is earned, not given by mistake.

But let's be clear: this isn't about becoming the best or most knowledgeable there is (an impossible and a rather lonely pursuit). It's about recognizing your own worth and contributions. It's about understanding that feeling like an impostor sometimes is part of the human experience, especially in fields as ever-changing and challenging as technology, programming, and cybersecurity. So, to those of you reading this, feeling like you're on the brink of being "found out," know this: you're not alone. We all feel it. We all battle it. And yes, we can all overcome it. Find what works for you, and when you do, embrace it, use it, and remember, the tech world is better for having you in it.

Branding Revisited

See, I promised we'd come back to this! 😊

Like the technology we're so passionate about, our personal brand is never static; it's a living, breathing entity that evolves as we do. If you've taken all of my stories to heart, adopt continuous learning as a badge, embrace networking, and you can overcome imposter syndrome, then you are truly ready to focus on the creation and growth of an exceptional personal

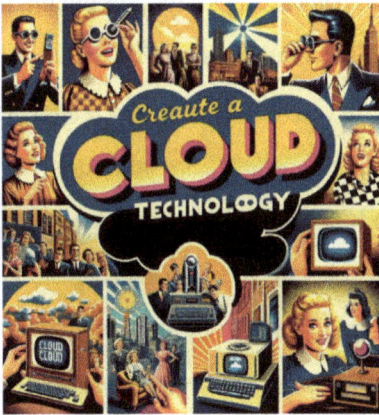

brand. The birth of your brand is both a moment of creation and a declaration of intent. It's the point where vision meets voice, where values crystallize into imagery and words that resonate.

This genesis is a deeply introspective journey, one where you chart a course based on your unique strengths, passions, and the value you bring to the digital frontier. This is not just about defining who you are to the world, but also about navigating your path with purpose and authenticity.

The bedrock of a compelling personal brand is a clear, powerful vision. This isn't about job titles or technical skills; it's about your larger mission in the tech landscape. What drives you? What change do you seek to make? Whether it's innovating in cybersecurity, democratizing technology, or anything in between, your vision is your beacon, guiding your journey and illuminating your brand.

Reflect on your career aspirations, the impacts you wish to have, and the legacy you aim to leave. This vision statement becomes your personal brand's North Star, guiding every decision and communication. If you want to be viewed as a thought leader, as a highly skilled and knowledgeable expert, as someone that others would kill for to be on a team, then make that your vision. Consider the qualities that define you, both as a professional and an individual. Are you a relentless problem-solver? A creative thinker? A collaborative leader? Your voice should echo these attributes, resonating with sincerity in every interaction. It will become the essence of your personal brand, ensuring that every tweet, post, presentation, or project genuinely reflects who you are and what you stand for.

Remember everything we discussed regarding building rapport, using soft skills, networking… this is your audience. A personal brand thrives on genuine connections like these. It's about engaging with your peers, mentors, mentees, and community in ways that foster mutual growth and understanding. Leverage platforms like LinkedIn, Twitter, or industry forums to share your insights, ask questions, and participate in conversations. Attend conferences, webinars, and meetups to expand your network and deepen your connections.

While personal branding is predominantly about intangible qualities, having a consistent visual identity can amplify your presence as well. A professional headshot, a unique logo or signature, or a consistent color scheme across your online profiles can make your personal brand more recognizable and memorable. Remember the first story about Scott and the lessons on building a great resume? Apply your brand to your resume and CV as well. Choose elements that reflect your professional demeanor and personality. Whether it's through a website, business cards, or social media profiles, ensure that your visual identity reinforces the essence of your personal brand.

Introducing your personal brand to the world is an ongoing process. It's not a one-time event but a continuous journey of demonstrating your expertise, sharing your experiences, and contributing to your field. Reflect regularly. Set aside time periodically to reflect on your journey. What new skills have you acquired? What lessons have you learned from recent challenges? How have your goals and passions shifted? Use these reflections to grow.

At its core, your brand is your story—a narrative that connects your past, present, and future. As you evolve, so should your story. Share your journey, the hurdles you've overcome, the triumphs you've celebrated, and the lessons you've learned along the way. This not only makes your brand more relatable but also more inspiring to those who might be walking a path similar to yours. Keep your finger on the pulse of the industry. What emerging technologies or methodologies are on the horizon? How might they impact your field or your role? By staying informed, you can anticipate changes and adapt what you share and how you interact to remain relevant and forward-thinking. Identify opportunities to showcase your expertise and insights. This could be through blogging, speaking engagements, contributing to open-source projects, or mentoring. Each of these activities not only elevates your visibility but also reinforces the value you offer.

Be open to feedback. Feedback is a gift, offering insights into how others perceive you and where there might be gaps between your self-perception and how you're seen by the world. Use this feedback as a guide to refine and adjust your brand. Branding is ultimately about introspection and outward expression, about finding your unique place in the tech world and filling it with confidence and authenticity in a way that truly captures who you are and why you're the go-to person for whatever your technical expertise may be. It's a reflection of your journey—ever-evolving, dynamically adapting, but always true to your core passions and values.

By laying down a strong, authentic foundation, you pave the way for a personal brand that not only stands out but also stands the test of time in the ever-changing tech landscape. And with a good brand and positive brand recognition comes opportunity to excel your career and your salary.

HEY YOU. YEAH YOU.
WE ALL START SOMEWHERE. YOU'RE GOING PLACES.

A Final Fable

In the tech crucible of Quantum Valley, amidst the silicon dreams and digital revolutions, Eliana Rivera carved a path illuminated by her unique vision and relentless drive. An entrepreneur and cloud architect specializing in Office 365 and Teams PBX solutions, Eliana's journey was a testament to the power of innovation and the indomitable spirit of those who dare to reshape the industry.

Growing up in a tight-knit family in the outskirts of Quantum Valley, Eliana Rivera was a curious child with an insatiable appetite for discovery. Her childhood home was a vibrant tapestry of Hispanic culture—filled with the sounds of lively music, the aroma of traditional dishes, and the warmth of familial love. Her parents, both artists in their own right, instilled in her a deep appreciation for creativity and the courage to pursue one's passions, even when they led down unconventional paths.

From a young age, Eliana exhibited a penchant for technology. Her first computer, a hand-me-down from an older cousin, became her canvas, her playground, and eventually, her window to the world. While her peers found solace in video games and social media, Eliana found hers in dismantling software, understanding systems, and piecing together the building blocks of what would become her future career.

Her family's gatherings often turned into impromptu tech support sessions, with Eliana at the helm, troubleshooting issues with the same ease and patience her grandmother used to solve puzzles. "You have a gift, mija," her grandmother would say, watching Eliana navigate through error messages and system settings with a determined gleam in her eye.

High school brought new challenges and opportunities. Eliana joined the robotics club, standing out not just for being one of the few girls but for her innovative designs and quick problem-solving skills. It was here, amidst circuits and code, that Eliana's dream began to crystallize. She wasn't content with just understanding technology; she wanted to use it to connect, to build, and to solve real-world problems.

Her decision to pursue a degree in Computer Science was met with mixed emotions from her family. Proud, yet anxious about the competitive and fast-paced world she was about to enter, they supported her, knowing that her passion and determination would see her through.

Quantum Valley's university was a melting pot of ideas and ambitions, and Eliana thrived. She absorbed knowledge like a sponge, excelled in her courses, and sought out internships that allowed her to apply her skills in meaningful ways. It was during one such internship that Eliana first encountered Office 365 and Teams PBX solutions, sparking an interest that would soon become her career's focus.

As graduation approached, Eliana knew that her journey was just beginning. Armed with her degree, little experience, and a burning desire to make a difference, she stepped into the professional world, ready to leave her mark on Quantum Valley and beyond.

This background laid the foundation for Eliana's quest to build a personal brand in Quantum Valley—a journey marked by the convergence of her heritage, her passion for technology, and her vision to transform the digital landscape. Her ambition? To democratize advanced cloud solutions, making them accessible to businesses of all sizes, and specifically targeting Hispanic businesses in the area (who she felt were underserved by the dime-a-dozen other consultancies in the area). Yet, as she navigated the complexities of her field, she knew that to truly make an impact, she needed to do more than just excel technically; she needed to build a brand, a voice that could cut through the digital noise.

Fresh out of college, Eliana quickly discovered that the world was not waiting with open arms for another cloud engineer, no matter how passionate or skilled. The first of her challenges was the sheer volume of competition. Quantum Valley was a magnet for talent, attracting the brightest minds from across the globe, each equipped with their own arsenal of certifications, experiences, and connections. Eliana, with her newly minted degree and a portfolio brimming with academic projects, found herself lost in a sea of candidates who boasted years of practical experience.

Networking events, once a source of excitement, became gauntlets of intimidation. At one such event, Eliana found herself sandwiched between a seasoned software engineer who casually mentioned their decade-long stint at a leading tech giant and a young entrepreneur who had just secured their second round of venture capital. "And what do you do?" they inquired, turning their attention to Eliana.

"I'm a cloud architect, specializing in Office 365 and Teams PBX solutions," she replied, trying to infuse her voice with confidence she didn't feel.

"Ah, a new graduate?" the software engineer asked, a hint of dismissiveness in their tone. The entrepreneur's smile was polite but fleeting, their interest visibly waning as they scanned the room for more established connections.

The challenge wasn't just the competition; it was also the pervasive underestimation and the subtle barriers that seemed to rise at every turn. Job applications went unanswered, and the few interviews that came her way ended with the all-too-familiar refrain, "We're looking for someone with more experience."

But perhaps the most disheartening obstacle was the struggle to gain visibility and credibility in a field that valued proven track records over potential. Eliana's attempts to reach out to potential clients for her consultancy often met with skepticism. "It's a risky investment, trusting someone so new to the industry," one potential client remarked during a consultation call, echoing the sentiment of many.

These challenges, each a roadblock on her path, began to weigh heavily on Eliana. Doubt crept in, whispering questions she had never thought to ask herself. Was her vision too ambitious?

Had she overestimated her ability to make a mark in such a competitive landscape?

As she sat in her apartment, surrounded by the very technology she aspired to master, Eliana couldn't help but feel a pang of isolation. The journey she had embarked on with such enthusiasm seemed to be faltering before it had truly begun. The dream of building a personal brand that could influence and inspire seemed increasingly out of reach, obscured by the realities of an industry that demanded not just skill, but evidence of impact and experience she had yet to accumulate.

Yet, beneath the comforting glow of her grandmother's kitchen, Eliana sought refuge from the storm of challenges she faced. Her grandmother, Abuela Rosa, with her wisdom woven through years of life's tapestries, was stirring a pot on the stove when Eliana began to share her burdens.

"Abuela, it's like I'm invisible," Eliana confessed, her voice tinged with frustration.

"No matter how hard I try, it feels like the doors of the industry just won't open for me."

Abuela Rosa listened intently, her eyes reflecting a mix of empathy and the spark of unyielded strength that had guided her through her own trials. Setting the wooden spoon aside, she took Eliana's hands in hers, grounding her with a presence that felt as enduring as the earth itself.

"Mija, remember the story of the cactus flower?" Abuela Rosa began, her voice steady and sure. "It blooms in the harshest conditions, not in spite of them. Just like you will. You have a gift, a vision, and a voice that needs to be heard. Perhaps it's not the doors you're knocking on but finding the ones that are waiting for your touch."

Eliana looked up, finding solace in her grandmother's words, a gentle reminder of the resilience that ran through her veins.

"And you have something special, Eliana. You speak the language of technology and the language of us and others. You can bridge worlds that others can't," Abuela Rosa continued, her eyes gleaming with an idea. "Why not start your own business? There are many small businesses in our neighborhood, owners and workers who might struggle with English but could flourish with someone like you guiding them."

Eliana's eyes widened at the suggestion, the gears in her mind turning. She had always envisioned herself making an impact on a large scale, not realizing the opportunities that lay in the very community that shaped her.

"Just the other day, I heard Señora Luisa complaining about her shop's internet and phone costs. Said she wishes there was someone who could help her understand it all, maybe find a better, less expensive way," Abuela Rosa added, a knowing smile on her lips. "Mija, you could be that someone. Use your skills to help our community thrive. Build your reputation from the ground up, one local business at a time."

The idea settled into Eliana's heart like a seed planted in fertile soil. Her grandmother was right. Here, in the heart of her community, were people who could benefit from her expertise, people who might welcome her assistance not with skepticism but with open arms. It was a perspective she hadn't considered, a path unexplored.

"Gracias, Abuela," Eliana said, a newfound determination lighting up her face. "Maybe I've been looking at this all wrong. It's time I make my own opportunities, starting right here, with the people and the businesses that are the lifeblood of our neighborhood."

As she hugged her grandmother, Eliana felt a weight lift off her shoulders. The road ahead was still uncertain, but now it was lined with the possibility of making a real difference, of building her personal brand in a way that was true to her roots and her passions. Abuela Rosa's advice had not only offered a solution but had rekindled the fire within Eliana, reminding her that sometimes, the most profound impacts are made not by conforming to the vastness of the industry but by nurturing the growth of the community around you.

Eliana embarked on her outreach within the community. Her first stop was Señora Luisa's shop, a cozy establishment known for its delightful array of local crafts and goods. Eliana stepped inside, greeted by the familiar chime of the door and the warm smile of Señora Luisa, who was arranging a display near the counter.

"Señora Luisa, I've heard you've been having some issues with your internet and phone services. I think I can help," Eliana began in fluent Spanish, bridging the gap with ease and warmth.

As Eliana explained her proposal to revolutionize Señora Luisa's communication setup through a tailored Office 365 Phone solution, the shop owner listened, intrigued and hopeful. Eliana's plan was not just about upgrading technology; it was about understanding and addressing the unique needs of her business, all while cutting unnecessary costs.

After securing Señora Luisa's trust and agreement, Eliana set to work. She liaised with the local ISP, negotiating the removal of an expensive business plan burdened with unnecessary static IP addresses in favor of a standard residential service line. The transition was seamless, spearheaded by Eliana's technical acumen and her unwavering dedication to her client's best interests.

The result was a significant monthly savings for Señora Luisa, who was overjoyed at the newfound efficiency and cost-effectiveness of her shop's operations. Eliana, in turn, secured a modest but meaningful monthly revenue for managing the solution, marking the successful launch of her consultancy.

Word of Eliana's success with Señora Luisa's shop spread quickly through the community. Before long, Señora Luisa herself became one of Eliana's most fervent advocates, urging her to visit another store down the street. "You must talk to José at the bakery," she insisted, her eyes alight with gratitude. "He's been lamenting his internet bill for months. Tell him I sent you."

Before Eliana left Señora Luisa's shop that day, she bought a beautiful piece of artwork and took a picture together. She left not just with a sense of accomplishment but with a deepened connection to her community. She had taken the first step in her journey, not by looking outward to the vast, impersonal expanse of the tech industry, but by focusing inward, on the very community that had shaped her. Each successful project, each satisfied client, was a brick in the foundation of her personal brand, built on trust, expertise, and a genuine desire to make a difference.

This was just the beginning for Eliana Rivera, whose journey to empower her community with cloud solutions was underway, one local business at a time. Inspired to continue and fueled by a renewed sense of purpose, Eliana set about continuing to turn her vision into reality. She started by crafting the digital and physical embodiments of her newfound venture. Night after night, she worked on creating a website that not only showcased her expertise in Office 365 and Teams PBX solutions but also resonated with the heart and soul of her community. The website was a blend of professionalism and personal touch, featuring not just the technical services she offered but also her story and her commitment to empowering local businesses.

Next, Eliana designed business cards that mirrored the aesthetic and ethos of her website. She chose colors that reflected both the vibrancy of her Hispanic heritage and the sleekness of modern technology. Each card bore her name, her services, and a simple yet powerful slogan: "Bridging Communities with Cloud Solutions."

An updated LinkedIn profile followed, detailing her journey, her expertise, and her unique value proposition. She positioned herself as not just another IT professional but as a community-focused consultant who understood the challenges and needs of local businesses.

She posted her pictures to the feed, and always made sure that her posts were written bilingually.

In the months that followed her initial success with Señora Luisa's shop, Eliana's consultancy flourished, transforming from a solitary venture into a beacon of hope and empowerment for the local Hispanic business community. Word of her expertise and her unique approach to leveraging technology for community growth spread far and wide, attracting a diverse clientele eager to embrace the digital revolution under her guidance.

Eliana's impact went beyond just the technical solutions she implemented. She became a symbol of what could be achieved when technology was married with a deep understanding and respect for the cultural and linguistic needs of a community. Her work helped bridge the digital divide, bringing state-of-the-art cloud solutions to businesses that had previously been sidelined by the mainstream tech narrative. And, importantly, she had built a unique brand for herself that differentiated her from everyone else.

As her reputation grew, so did her team. Eliana was joined by other like-minded professionals, each bringing their own skills and passions to the venture. Together, they expanded the consultancy's offerings, always with an eye toward how they could serve not just the business needs but also the broader aspirations of the community. Workshops, free training sessions, and mentorship programs became as much a part of their services as cloud architecture and digital transformation projects.

Eliana's journey from a solitary entrepreneur to a community leader underscores a powerful truth: that success in the tech industry is not measured solely by knowledge or skills but by the people we know, the values we represent, and the impact one can have on the lives of others.

I hope you loved reading this story, and all the fables throughout the book, as much as I loved writing them. I think it's very apropos that this is the final fable as we wind down our journey together. Read it through the lens of everything that has come before it.

As you reflect on everything, consider your career as one of those long, sprawling TV series everyone seems to be talking about. Early seasons (a.k.a. your first job or project) might feel like they're moving at a snail's pace, filled with character development (your skills) and plot twists (unexpected job challenges). But it's the later seasons, where the long arcs pay off, that truly define the series. Similarly, your career is about the long game; it's about accumulating experiences, learning from failures (yes, even that project that went up in digital flames), and gradually moving towards your ultimate goals.

Remember, in the tech industry, being an overnight success is as rare as finding someone who actually reads the Terms and Conditions. So, take a breath, pace yourself, and remember that every troubleshooting challenge solved, every client won, and every project completed is a step towards your season finale.

Now, I don't want to sound like one of those motivational posters featuring a soaring eagle and a vaguely inspiring quote, but here's the deal: the only person who can write your epic career saga is you. Whether you're dreaming of launching your own startup, eyeing that senior developer position, or wanting to master the latest in cybersecurity, it's time to take action.

Let's break it down into 'season arcs':

Season 1: The Quest for Knowledge. Upskill, reskill, treadmill (for those who think and learn best on the move). Learning never stops in the tech world, so embrace it. Online courses, workshops, conferences—immerse yourself. And remember, tech skills are important, but soft skills are too!

Season 2: The Networking Chronicles. Connect with other professionals, not just on LinkedIn, but in real life too. Remember, the tech community is your tribe, and there's strength (and opportunities) in numbers. There is an ocean of opportunity out there and it's often related to who you know, not what you know.

Season 3: The Rise of the Brand. Become something unique or stand-out. Don't get lost in a sea of look-alikes with poorly written resumes and painful social interactions. Be confident, authentic, and ensure your digital presence is on point. You can't just be the part in your heart, you have to look the part as well.

Season 4: The Leap of Faith. Apply for that job. Launch that startup. Present your idea to the team. Yes, it's scary. Yes, you might fail. But remember, every giant started as a dreamer with a crazy idea. The only way to grow your career and your salary is to take those leaps.

The Penultimate Episode: Crafting a Legacy. Remember that your career in tech is a story still being written. There will be plot twists, cliffhangers, and maybe a spin-off or two. And with each challenge faced and each achievement unlocked, you're not just building a career; you're crafting a legacy for yourself. Your finale has yet to be written, your swan song is yet to come. And so, I cannot write that for you, only you can write it for yourself.

As you stand on the precipice of your next big adventure, remember our journey together and all of the fables, all of the lessons. Remember the power of perseverance, community, and innovation. Then, take a deep breath, and jump into your own epic story. And who knows? Perhaps one day, someone will be reading your saga, inspired to start a journey of their own.

Afterword

Wow! I can't believe I'm here, in this moment, with you. It almost feels like time traveling to me as I think about it – a piece of my past that has become your present, in hopes to help you with building your future. I am so incredibly humbled to have you reading this, to have been able to share my very first book with you. It is truly my hope that you have found the advice, approach, stories, jokes, memes, and everything I've poured into this book useful and meaningful to you.

I mentioned earlier that I was an avid tabletop role player. Writing is not new to me, but writing a book is (especially a nonfiction one). Sure, I've written technical manuals and how-to guides, but never something like this. I also find writing fiction much more comfortable and interesting in general, which is one of the reasons I wanted to incorporate modern-day fables, as kitschy and non-epic as they may be. And now, after a year, a ridiculous number of edits, and far too much caffeine... it's done. I did it. And if you're still reading this, I hope it means I did it right.

I do plan on writing more and publishing future works. While this book was focused on individual tech professionals and how they can grow their career, I think it's equally as important to help non-technical people, particularly small businesses and non-profits as well. As much as I want to help as many techs excel as possible, I also want to help organizations choose qualified, worthwhile talent and partners to hire and invest in. In a way, that makes following the advice in this book much more important. I am also very interested in making information security more accessible and more organic in our everyday lives and considerations. Who knows, maybe I'll even write an epic space opera or high fantasy novel!

Since I work full time, I had to work on this outside business hours, on my own personal time and weekends. Because of this, I feel it's crucial to share one final piece of parting advice and touch upon a shadow that often trails the steps of those in high-demand, high-stress fields like ours: **burnout**. It's the silent system failure many of us don't see coming until the screen of our resolve goes blue, and the reboot button seems frustratingly out of reach. Burnout doesn't discriminate; it can dim the brightest minds and dull the sharpest talents, leaving us questioning our passion and our path. And if you're on a journey to supercharge your career, it will become even more likely.

So, my parting advice to you is this: guard your flame. Passion is a renewable resource, but only if you nurture it. Find balance, seek joy outside the command line and code, and remember that it's okay to step away and breathe. The digital world will spin on, and you'll return with a clearer mind and a stronger spirit. Plan time off, don't wait until you're sick and need a mental health day. Plan in advance. Remember what you're working for and why. Work to live, don't live to work. None of us are immune to the pressures that come with ambition and achievement. It's also paramount to have a support system. Whether it's family, friends, colleagues, or even a loyal pet, these are the connections that ground us, remind us of our worth beyond our work, and help us through the inevitable tough patches. I recognize that this is all easier said than done, and in no way am I suggesting it will be easy for you. Any fears and trepidations about your potential workload are valid. But something one of my high school teachers said to me has stuck with me throughout the years (thank you, Mr. Bratowicz). "You can take it easy now, and life will be hard. Or you can work hard now, and life will be easier." Of course, I didn't listen to him at the time… (what snot-nosed jerk teenager ever listens to good advice?) but the words ring true now. The more time and effort you put into something upfront, the easier it will be in the long run. This doesn't make the decision to start any easier, nor does it mean you won't have setbacks, but I hope it helps justify why it's important to take the leaps even though they will be hard.

In my journey, I've been very blessed to have empathetic and understanding family members and friends who have been extremely supportive in my journey. I'd like to specifically thank another high school teacher of mine, Mrs. Hopper (perhaps better known as Kristen Houghton). She not only stands out in memory as one of my favorite teachers, but also as someone who reminds me that it's never too late to pursue something new. She's the accomplished author of the Cate Harlow detective series and has been an incredible help to me with regard to actually getting published!

Regarding my personal support system, I want to extend a special 'thank you' to my Uncle Karl Ross and Aunt Nancy for their love, support, generosity, and kindness all throughout my life, even when I may not have been the easiest person to care about.

Finally, I cannot possibly write enough to truly encapsulate the depth of love, admiration, and appreciation I have for Meagan Vidal, my proverbial partner in crime and definitely my better half. Within the abyss of my climate-controlled beating CPU, underneath an avalanche of what I like to assume are all logical ones and zeros, there is a twinge of undeniable emotion that has grown at least three sizes too big for its chamber. I know I don't often express it to her poetically, frequently, or with particularly overt grandeur, but it is a profound, if not magical sensation to know that not only is someone capable of such love and support, but they truly and deeply make you a better person just for having known them. And to Meagan's daughter Josephine, whom I hope will one day be able to read and understand this... you are the warmth and ever-present burning of the golden sun on even the coldest of winter days. I love you both.

And now, the only thing left to do...

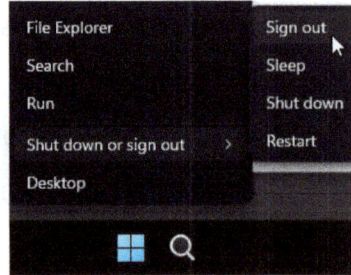

File Explorer	Sign out
Search	Sleep
Run	Shut down
Shut down or sign out >	Restart
Desktop